Aristotle's

On the Soul

On Memory and Recollection

Aristotle's

On the Soul

and

On Memory and Recollection

Translated by Joe Sachs

Green Lion Press
Santa Fe, New Mexico

Manufactured in the United States of America.

Fifth printing, 2019.

Printed and bound by Sheridan Books, Inc., Ann Arbor, Michigan.

Published by Green Lion Press
www.greenlion.com

Green Lion Press books are printed on acid-free paper. Both softbound and clothbound editions have sewn bindings designed to lie flat and allow heavy use by students and researchers. Clothbound editions meet the guidelines for permanence and durability of the Committee on Production Guidelines for Book Longevity of the Council on Library Resources.

Cover and dust jacket design by Dana Densmore and William H. Donahue with help from Nadine Shea. Cover photograph of snow monkey in Japanese thermal pool by Tom Brakefield; used with permission of Bruce Coleman Inc.

Cataloging-in-Publication Data:

Aristotle
On the Soul / by Aristotle
translated by Joe Sachs.

Complete English translation of Aristotle's
On the Soul and On Memory and Recollection,
with introduction, footnotes, glossaries, bibliography, and index.

ISBN-13: 978-1-888009-16-3 (sewn clothbound);
ISBN-13: 978-1-888009-17-0 (sewn softcover)

1. Aristotle, works, English. 2. Biology. 3. Philosophy. 4. Classics.
5. History of Science. 6. Psychology—early works to 1850.
7. Soul—early works to 1800.
I. Aristotle (384–322 BCE). II.Sachs, Joe 1946– . III. Title.

B415.A5S232001

Library of Congress Catalog Number 2001-091953

Contents

1. All individual chapter titles are supplied by the translator.

The Green Lion's Preface

The Green Lion Press is delighted and honored to be presenting Joe Sachs's translations of Aristotle's *On the Soul* and *On Memory and Recollection.*

On the Soul addresses the question of what it means to be alive. Closely related thematically to the *Physics*, it is Aristotle's application specifically to living things of his scrutiny of the self-organization of all natural things. Aristotle argues that life is not just the result of an arrangement of parts, nor can it be adequately understood by positing a separate soul that is imposed on otherwise brute matter. Though thousands of years old, Aristotle's penetrating analysis explains important aspects of our actual human experience of our embodied selves, and the world with which we interact, that present-day materialistic or dualistic mind/body theories cannot.

On the Soul also contains Aristotle's fascinating account of light and colors. Rejecting the ray theories of contemporary mathematical optics (with which he was familiar), he argued that we see things because light activates the latent transparency of the medium between our eye and a visible object. Although this view had many rivals throughout antiquity and the Middle Ages, it remained a strong contender until the early seventeenth century.

On Memory and Recollection is a little gem of a book that considers how it is that we remember things, and how we recollect something that temporarily escapes our consciousness. It extends the study of the thinking soul that was initiated in *On the Soul.*

In these new translations, Joe Sachs continues his groundbreaking project of restoring Aristotle's thought to its original freshness, sweeping away the cobwebs of scholastic terminology. Gone are the abstract Latin-based words, replaced by a radically new rendering, in ordinary English, of words that were, after all, everyday Greek. The result is a version that will reward the careful reader with a new appreciation of the power and subtlety of Aristotle's reasoning.

Joe Sachs's brilliant introduction—exquisitely nuanced and insightful, delighting the reader further with his characteristic dry, understated, but precisely-focused wit—makes Aristotle's understanding of the soul luminously clear. Sachs opens this understanding to us by beautifully (and movingly) expounding soul in Aristotle's own terms and by contrasting Aristotle's view with more modern conceptions of soul. These latter are due primarily to the efforts of Descartes to begin philosophical investigations with, and limit them to, what is clear and distinct—that is, to the surfaces of things. Sachs draws out the profound consequences of this for our view of ourselves and our relation to the world, consequences that diminish the possibilities of richness and sense of meaningfulness of both.

As with our edition of Aristotle's *Metaphysics,* we have designed this book to be easy to read, study, teach with, and use as a basis for discussion. Having ourselves read, studied, taught, and discussed this text with students and colleagues over many years, we know what we as users needed and wanted in presentation and layout, and we have taken pains to provide those things.

As readers we wanted a sewn binding for the softcover edition so that pages would not fall out even on repeated readings. We wanted high quality paper and generous margins for making notes. We wanted a sturdy cover that would not curl or fray when the book was carted around in book bag or student backpack. We wanted a glossary that thoroughly explained translation choices. We wanted footnotes, not end notes, so that we could see at a glance what the note said and easily either read it or defer attending to it without breaking the flow of our reading of Aristotle. We wanted the footnote number in the text to be large or bold enough that it could be found if later we came back to a footnote and wanted to find the place in the text where the footnote was invoked. We wanted Bekker page number and line number ranges,[1] when cited, to be specified in full so as not to be ambiguous or misleading.

As scholars, we wanted all those features and other things as well. We wanted the Bekker numbers and the Book and Chapter numbers to be easy to find and follow. We wanted a glossary of the Greek terms, in Greek alphabetical order, as well as a glossary of the English terms.

As teachers and participants in discussions of the text, engaged with others using other translations, we wanted all that and in addition anything and everything to let us quickly and accurately find the place in the text to which someone was referring. We wanted to be able to use the book in a seminar or lecture question period and be able to find a place mentioned easily enough that we didn't have to take our attention away from the discussion and put it into the hunt.

For the occasions when an interlocutor made reference to a bit of text by Bekker page and line numbers, we wanted those numbers in the margins, not embedded so discreetly in the text that repeated scanning of the page failed to find them. We wanted every one of those marginal references spelled out in full, repeating the citation of the page number with every line number, so that we didn't have to turn pages back or ahead to find out on what page the orphan line numbers we were looking at appeared.

For the occasions when an interlocutor made reference to a bit of text by Book and Chapter numbers, we wanted both in the running heads on every page so that it could be turned to

1. Page, column, and line numbers of the 1831 Bekker edition of the Greek text of Aristotle's works.

easily—without having to do as we have done and take each new translation or edition of Aristotle and go through writing those things in by hand on every page.

All these things are provided here. As with our *Metaphysics* edition, we have made the book we always wished we had when we read and used Aristotle, and we hope that these design features will please you and serve you as well as they will us and that they will make your engagement with Aristotle direct and without external impediments.

Dana Densmore and William H. Donahue
for the Green Lion Press

> The mule driver takes for granted the performance of his animals. He assumes that they can see and in seeing they apprehend the peculiarity of the road and behave accordingly. Yet he is only a simple man.
>
> —*Erwin Straus*

> *Socrates*: If I don't know Phaedrus, I've forgotten myself.

Introduction

Recognizing the Soul

To the simple and unsophisticated among us, or perhaps to all of us at unguarded moments, there is a certain comfort in the fact that we share the world with so many varieties of beings that are so much like ourselves. To commune with nature means to recognize something in ourselves by its reflection in something outside us. We do this when we stare into a fire, when we listen to waves breaking on a shore, or when we drink in with our eyes any unspoiled formation of the earth. The peacefulness that comes when such things absorb us is a way of being at home, of discovering that we have a place. But the same sort of peaceful rest is enhanced by the still greater kinship we feel in being among plants and trees, and especially in undemanding encounters with animals. They recall us to what we are; they remind us that we have souls.

I believe that I am here using the word *soul* in exactly the sense that Aristotle uses the word *psuchē*. Much can be written and read about the history of these words, and the connotations each has collected by association with various beliefs that have come and gone in the course of time, but there is a primary meaning at the root of each. That meaning precedes anything that can be read or written, and is found only in experience. I have tried to evoke that experience in the preceding paragraph. When Aristotle sets out to define the soul, he is not saying "let us agree to use this word this way," but making a step toward understanding our common experience of an aspect of the world. The inquiry can go nowhere if we do not have in common the thing inquired about. Do we?

In Book II, Chapter 1, of his *Physics,* Aristotle says it would be ridiculous to try to show or convince anyone that there is such a thing as nature. Anyone in doubt over such a thing could not participate in an inquiry, which requires at least that one distinguish between what is known directly and what is known through something else. We could go through the motions of inquiring, like blind men reasoning about colors, making statements about words, but with insight into nothing (193a 3–9). The topic of nature is entirely analogous to that of soul in this respect. More than that, it is almost the identical topic. The word for soul hardly occurs at all in the *Physics,* and plays no important role there, but that is because its meaning is absorbed into that of nature. Nature, in Aristotle's view, is always *the* nature of some kind of being, to which it belongs primarily and in virtue of itself (192b 22), and almost all such beings are living things. In the *Physics,* though, that *almost* is crucial, as nature unfolds itself between two poles. Nature is predominantly living nature, but it is also the ordered cosmos that sustains life, and ensures its possibility. The focus of the *Physics* is on what is shared by living things and the elemental universe, but the recognition of nature that the inquiry presupposes must include, and cannot be present without, the recognition of soul.

But what kind of recognition is this? Is it, for instance, clear and distinct? If we are vague about what constitutes the soul, and apt to confuse it with other things, it may be that it is nothing at all, or only some sort of illusion. But once more the *Physics* has advice to offer us. Book I, Chapter 1, of that work tells us that it would be unnatural to begin an inquiry with anything clear and distinct. The place to begin is with what is familiar, which is bound to be confused and indistinct. Clarity and distinctness prematurely arrived at might be imposed on things, or invented by us in our impatience. The clarity and distinctness that are worth attaining are those that emerge through the honest examination of what we encounter in experience. And the experiences most worth examining might be just exactly those that have some mystery about them, some hidden depth.

What would it mean for something to be experienced as clear and distinct from the beginning, from first glance? Would it not mean that it belonged only to the surface of things? The words "clear and distinct" are associated especially with the seventeenth century philosopher Rene Descartes, and in his use of them, they have only this narrow meaning.[1] In the sixth chapter of his book *Le Monde*, one of the founding documents of modern mathematical physics, Descartes goes so far as to deny that anything in the world has any qualities. This is an imaginary world, created by Descartes himself for the sake of perfect knowability, but finally identified with the world we dwell in. It contains nothing that can, even for a moment, resist being fully and clearly known. The only kind of form that the matter of this world can have is shape, and besides shape it can have nothing at all but size, position, direction, and speed.

1. There is something both right and wrong in allowing Descartes to have proprietary rights in clarity and distinctness, at which nearly all philosophers aim and of which most speak. By elevating them above other aims, and attempting to shortcut the road to them, he stakes everything on their power, and risks everything they cannot deliver.

Lacking all qualities, it has only quantities, and everything that is true of it is spread out right on its surface.

This turn toward the superficial is motivated, in part, by the lure of minimizing the human role in our own knowing. Despite the fact that instruments of measurement must be designed, interpreted, applied, and read, the quantities they register seem to stand beyond all doubt, and eliminate any need for trust. While quantities become accessible to us only through perception, they seem to leave their perceptual heritage behind. Qualities, on the other hand, not only must be perceived; they are nothing at all outside the act of perception. A wavelength can be registered on a machine, but blue or red can only be experienced. There is a gulf between quantity and quality that can never be bridged. Listen to Section 17 of the *Monadology* of Gottfried Leibniz, a powerful critic of Descartes:

> It must be confessed that *perception* and that which depends upon it are *inexplicable on mechanical grounds*, that is to say, by means of figures and motions. And supposing there were a machine, so constructed as to think, feel, and have perception, it might be conceived as increased in size, while keeping the same proportions, so that one might go into it as into a mill. That being so, we should, on examining its interior, find only parts which work one upon another, and never anything by which to explain a perception. Thus it is in a simple substance, and not in a compound or in a machine, that perception must be sought for. Further, nothing but this (namely perceptions and their changes) can be found in a simple substance. It is also in this alone that all the *internal activities* of simple substances can consist.

This is a classic statement, in the language of the modern world, of the irreducible and unarguable necessity to acknowledge the presence in the world of souls. That inward depth of life that opens in perception cannot be found on or between the extended surfaces of bodies, and when we cut bodies open, all we find on *their* insides are more outsides. This is an ultimate distinction discovered by philosophic reflection. Even Descartes does not deny the distinction. He merely restricts it, hoards it, jealously denies its presence anywhere but in himself. Since the evidence of soul must be clear and distinct from the outset, there is no soul but one's own. One may grant, graciously or grudgingly, that other humans also perceive, since the practical necessities of life would make that denial absurd, but the animals are mere machines, devoid of any inward life.

This restriction in the distribution of souls involves a decisive shift in the meaning of soul. Life remains in certain beings, dogs or horses for example, that have no souls. The inward depth required for any sort of perception is granted only to beings that talk and understand, or read and write books. Since souls now belong only to things that think, the meaning of soul contracts to become what we now call mind. For Descartes, when he writes Latin, this is a shift only in word endings, from *anima* to *animus*. For each of us, it is a shift of ourselves out of the world. If it is not perception, but merely some mechanical response to stimuli, that guides a mule to walk along a mountain path, then I need not understand my own walking as guided by perception either. My body too is a machine, working on its own, while I dwell elsewhere, out of the world, inside my own mind. David Hume, in his *Treatise of Human Nature* (Book I, Part 4, Chapter 6), tells us "The mind is a kind of theater, where several perceptions successively make their appearance; pass, re-pass, glide away, and mingle in an infinite variety of postures and situations." The inwardness that might be seen as an enrichment of the world becomes, in isolation from the world, an impoverishment of ourselves. The predominant philosophic and scientific tradition of the last four hundred years has taken away our souls.

Now from the point of view of the advice Aristotle gave

us about how to begin inquiring, this abandonment of the soul is a consequence of the premature desire for clarity and distinctness. In our primary experience of things, the evidence of what we are and what other things are reflects back and forth. We discover ourselves not in solitary self-examination, but through our recognition of living things. If a bird lands on a branch near us and cocks its head in our direction, chances are we will smile and say "hello." This is sometimes described, dismissively, as anthropomorphism, sometimes, with more approval, as empathy. Both descriptions seem off base to me. Both imply that I already know what I am; they disagree about whether I am entitled to project that known nature onto the little bundle of bones and feathers in front of me. But is it not possible that this unguarded encounter might be of a more original significance? Might not the bird be one of many mirrors in and through which I begin to encounter myself? This is not an occasion of certainty or clarity, of deduction or hypothesis, but of an obscure inner accord. In it, inner and outer are not the severed distinct halves of a mind-body problem, but intertwined aspects of a single recognition. I see the bird see me. I believe the bird perceives me, not because I have a theory to explain it, but because it makes immediate sense. I have a soul because I recognize a fellow soul; it takes one to know one.

Our encounter with other living things is mutual in a number of ways. The evidence that animals have souls is not projected onto them, but drawn out of us. The meaning of their life is not found by scaling down our own, but by rising to meet theirs. The inner life of the animal presents itself to us in its outer activity, and teaches us that we too dwell innately in our bodies. When the bird flies away, and doesn't bump into the branches and other obstacles it glides around, we recognize two things: that the bird itself originated its motion, and that it guides its course by means of perceiving the things around it. Living motion reflects an inner source, just as living perception requires an inner awareness. Life is a two-way relation with the things around us, never distinct from the body, never merely the performance of the body. The living body is centered inwardly as one perceiving being, and centered outwardly as one source of motion. And just as its perceiving guides its motion, its motion guides its perceiving: the bird cocks its head for a better look; we reach out to feel something we see, or walk around it for a more rounded view. The perceiving center *is* the moving source and vice versa. The two unities in the experience of the living body are one and the same, one soul.

Notice how this way of grasping things differs from anthropomorphism or empathy. If I am a self-enclosed mind, clearly known and clearly distinct from body, the only way to get to anything else like myself is to project it. If the world external to me is extended body, clearly known and clearly lacking everything that I don't clearly know, the only way to make it resemble me is to distort it. But these are highly artificial assumptions to bring to the world, and certainly not necessary ones. A native responsiveness to the things around me shows me my reflection in a thousand ways as more than mind, just as those things show themselves as more than body. I can't anthropomorphize as long as I am still unclear about the *anthropos,* nor empathize with something that is still teaching me what I am. I can recognize things like me, and grow into that recognition. I may not have achieved clarity, but I may instead have a chance to achieve depth. That depth of understanding, which does not surrender the depth of life perceived within or without, may keep me at home in the world.

There are then two major ways to go in thinking about everything there is. The way adopted for the most part in recent centuries, not in practice but in the most approved kinds of theory, has been to posit a picture of the world that excludes souls, and to try to cope somehow with the wreckage. The other way, seen more clearly by Aristotle than by most other thinkers, is to realize that the world must be so constituted in the first place that soul and the activities of life are genuine possibilities within it. I have mentioned already that the word for soul is nearly absent from Aristotle's *Physics* because its meaning is already built into the topic of that work, the conception of nature as a source of motion and rest within beings that have such sources in their own right. Even in Aristotle's inquiry into the *Parts of Animals*, the word for soul is very seldom used, but there it is equated with form, *eidos* (641a 17). This is in accord with what is learned in the *Physics*, since there the conclusion reached in Book II, Chapter 1, where nature is first characterized, is that the nature of anything is its form, *morphē* or *eidos* being used interchangeably there. In the *Metaphysics*, where the whole constitution of everything there is becomes the topic of inquiry, the first step of Aristotle's main argument is to trace all being back to things that are intact and independent. The primary examples are animals and plants (1028b 9–10), and the source responsible for the thinghood of these things is again identified as form (1029a 29–32), again with *morphē* and *eidos* used interchangeably. The structure of this world, not some other that we may imagine or invent, but this one that we experience, must accommodate the structure of the natural and that means, pre-eminently, the living.

What is that structure? Aristotle traces it down many roads. In the Platonic dialogues, Socrates is always asking the question *ti esti*?, what is it?, about any single look that makes many perceptible things all be the same one thing. It is possible to treat this as a logical question, having to do with classifying the world, attaching universal names to particular things. But we can sort the world into any classes we please. Aristotle is interested in the way the world sorts itself out, and this is visible in anything that keeps on being the same while constantly undergoing change. And what if that enduring sameness is of a kind that could not even be present except in the course of change? In that case, the thing does not hold out passively against change, but absorbs change into itself, molds it into a new kind of identity, a second level of sameness, a higher order of being. For such a being, to be at all depends on its keeping on being what it is. Aristotle sums up this way of being in his phrase *to ti ēn einai,* for what this sort of thing is cannot be given by some arbitrary classification of it, but is what it keeps on being in order to be at all. Its very being is activity, being-at-work, for which his word is *energeia,* and because it is a wholeness of identity achieved in and through being-at-work, Aristotle invents as a name for it the word *entelecheia,* being-at-work-staying-itself. That is the depth that he discovers within the notion of form, by a dialectical examination and gradual clarification of the originally obscure experience that the world presents to us things of various looks.[2]

The understanding of the form of anything as its being-at-work-staying-itself is a fundamental explanatory structure, and it is clear that life falls under it. For anything, from plants on up, that grows into a mature form, and maintains an organized body by appropriating material from its surroundings, to be at all is to keep on being what it is by being and staying at work; the cessation of such activity is death, and a dead oak tree, though it is still something, is no longer an oak tree. Its being was its life. Now two things may be noticed at this point. First, while this conception of being-at-work-staying-itself applies to all living things, it seems precisely to miss the two characteristics that gave us evidence of souls, perception and self-motion. By fitting so well the vegetative level of activity, that of nutrition and metabolism, it leaves the mystery surrounding the higher life of animals. But second,

---

2. For the identification of form with *entelecheia* and *energeia,* see especially *Physics* 193b 3–8 and *Metaphysics* 1050b 2–3.

we may notice that being-at-work-staying-itself is not a mere synonym for life. It is an intelligible structure that might apply to things other than animals and plants. Once it is available, philosophic reflection might be able to make headway along roads other than that on which it first came to sight.

For example, imagine a body moving perpetually in a circular orbit. As a moving body it is always going from one place to another, but this does not capture what it is doing. Every place it moves away from is just as much a place it is moving toward. The from which/to which distinction breaks down in this case. An orbit has not been grasped as long as it seems to be many places; it is one place where the orbiting body always is. This is the same structure we have described as being-at-work-staying-itself, in a very simple instance. The motion of the orbiting body is molded into a higher kind of stillness. That stillness is never visible to the eye, but it is the true and invisible look of the body's motion, a unity grasped through perception by means of thinking, a constancy present amid change because it is the changing that is constantly achieving it. This example is taken from Book VIII of Aristotle's *Physics*, and the idea of regular circular motion was important to astronomers long before Aristotle's time, but whether you think the sun moves or the earth, in a circle or an ellipse or any other closed path, or that the truly constant motion belongs to electrons or other parts of an atomic structure, wherever you find a constant pattern of motion, you will be in touch with evidence of the nature of things, and Aristotle's analysis of it will apply. Any type of physics, dealing with moving and changing things that are not alive, might still uncover something that is at work staying itself.

So the structure that matches up with that of a living thing is also found apart from living things. It may also be found in particular activities within living things. An example is discussed in Book VII of the *Nicomachean Ethics*. Aristotle considers there the opinion of some people that a pleasure is a perceptible motion by which some good condition comes about in the body. On a hot day you become dehydrated; as you drink a glass of water, the proper balance of moisture is restored to the body. An intense pleasure is present, but ceases when the good condition is achieved. Pleasure is then a good thing of a secondary kind, a sign that the true good is on its way. If we had never become dehydrated in the first place, we would have been better off, spared both the pleasure and the pain; or, from another point of view, if pleasure is our aim, we should cultivate it by deliberately inducing the painful condition. Aristotle considers all of this a misunderstanding (1153a 8–17). A pleasure is not a motion but a being-at-work, completely and exactly what it is at every instant that it is present. Also, according to Aristotle, every way in which we are simply at work, when it is unimpeded, is a pleasure.

Clearly, a lot depends on what sort of thing a pleasure is. Just being able to pose the question whether it is a motion or a being-at-work opens the possibility of a new depth of understanding. But it is not just a matter of seeing alternative explanations. Adopting one opinion or the other will make a difference in life choices, and in the possibility of happiness. If pleasure is a motion out of some disturbed condition, it is always in conflict with a neutral state of healthy activity, and achieving pleasure must be a matter of compromise and trade-offs with one's own well-being. If pleasure is a being-at-work, it might be in complete harmony with one's natural state of well-being, there to be settled into if only false opinions and a childish state of distractedness stop getting in the way.[3] The highest human condition of happiness would then have the same structure as the lowest vegetative condition of mere

3. These conclusions are drawn in Book VII, Chapter 3 of the *Physics*.

life, differing from it in content, but in no way in conflict with it.

Aristotle's formulation of the idea of being-at-work might then be a liberating thought, even about things like pleasure that we seem to be intimately acquainted with. Let us see if it offers any help in thinking about perception. The occasion of a perception is some sort of interaction between the bodies around us and our own body, but the bodily event does not constitute the perception. Poking a rock with a stick doesn't seem to cause it any pain, and opening the shutter of a camera doesn't enable it to see anything. Likewise, whatever we think goes on in our sense organs and nervous system still just gives us a model of a succession of bodily interactions and events. A description of electrons flowing from the retina along neural pathways does not differ in kind from a description of a complex camera or even of a pinball machine. It is still just Leibniz's mill, in which nothing that happens can explain a perception. First of all, as Leibniz said, the perception is nowhere in the complexity of parts external to parts; it can only belong to something that is one and simple. This begins to suggest something analogous in structure to our circular orbit.

A whole series of bodily events may be necessary to allow us to have the simplest perception, say of a note sounded on a piano. But this multiplicity of organs and motions does not determine the structure of what is perceived. The unity of the tone heard is a purely perceptual unity that does not reflect the organization of the body parts that allow it to be heard. Parts of the body do move and act against one another, but the experience of hearing the tone seems to correspond to a single way of being-at-work that belongs to the body as a whole. The external event that produces the tone in some way imposes the unity of its way of being-at-work upon certain parts of our bodies. The way Aristotle puts this, at the beginning of

Book II, Chapter 12, of *On the Soul* is that, in perceiving, the form of the perceptible thing is received without its material. Among Aristotle's predecessors, the only attempt to explain perception was by saying that the soul contains all the material elements of the surrounding world, so as to respond to like by like. (See Book I, Chapter 5.) But we do not need to have in us the metal of the piano string or the wood of its sounding board in order to hear it, but rather something with the power to be at work as they are at work.

And even this is not sufficient. We have traced a first level at which the perceiving body must take on not the mere motions of the things around it, but rather something corresponding to orbits, higher order wholes of being-at-work. But this capacity can also be found in the parts of a microphone, and again in those of a loudspeaker, or, as Aristotle points out, in the air (424b 14–18), which permits these things to become intermediate links in the process of hearing, but does not make it possible for them to hear. We have spoken above of the inwardness of the act of perceiving, but this metaphor, which we borrowed primarily from Leibniz, may be less fruitful than his distinction of the simple and the compound. In the body, a multiplicity of motions is molded into a single way of being-at-work, but again within our perceptions themselves there is a single awareness, a second forming of a third level of activity, the being-at-work of the perceiving soul. This is perhaps what Aristotle means when he says that the intellect is a form of forms (432a 2). Since form is being-at-work, a form of forms is a higher unity, a hierarchical structure organizing change at two levels. Just as the organs of the living body are wholes subordinated to a single comprehensive whole, so too the items of perception are wholes of activity comprehended into the single activity of perceiving them. In order for there to be perception at all, some one act must judge that what is sweet is not white (426b 12–23), so that perception is already on the borderline of thinking (429b 14–21, 431a 20–22, 432a 30–31).

Thus, an adequate description of perception must go beyond material to form and then to form again, beyond motion to being-at-work and then to a second level of being-at-work. Is this an explanation of perception? It is not an account of the physiology of the sense organs and nervous system, but it is an account that shows why any such physiology must fail to explain perception. If we mean by explanation the reduction of something to something else more familiar and easy to grasp, then many things will have no explanations. For example, suppose you want an education. Why do you want it? If it is to get a good job, or to please your parents, or to feel superior to those who are not educated, or out of curiosity of some sort, then it is easy to explain. But if there is something about learning that none of these reasons addresses, something in you that craves to be at work, some sort of knowing that you seek without guarantees that it exists and for no reason other than for its own sake, then the best explanation available is a description that is adequate and truthful. It is likely that, no matter how much you reflect on what you are seeking, some obscurity will remain in your conception of it. You are stretching yourself to understand something higher, not reducing it to something lower and already understood.

The Cartesian claim that animals are machines can lead two ways. It can be taken as reducing animals to something less than ourselves that we know how to analyze exhaustively, or it can lead to the decision that we ourselves are nothing but machines either. The latter choice also has two branches, since it can mean that perception is some sort of illusion that accompanies a mechanical reality, or that perceiving itself has been thoroughly explained when all the events in the sense organs and nervous system have been described. Of these two beliefs, each is in some way self-contradictory. If I say that vision is an illusory experience that occurs when light waves stimulate a flow of electrons from the retina, what do I mean by the word retina? Is it not something that is known by perception? As long as this sort of explanation is doing pure mathematics with an imaginary world, it is safe from objection, but as soon as it is applied to the world we experience, it becomes subject to the testimony of perception. The retina is not a mathematical construct, but a representation of a body with attributes and actions known only by perceiving them. The truth of observation and experiment is that it is rooted in perception, and it cannot then erase its own source by concluding that the thing perceived is not a thing perceived. The other version of this misunderstanding is the one that does not dismiss perception as illusory, but says that perception is nothing but a set of mechanical processes in the body. This is naked self-contradiction, the mere substitution of one word for a group of others that exclude its meaning. It is the same contradiction that is exposed by Leibniz's example of the mill, or by asking whether a camera sees, a contradiction discovered at a primitive level of experience that cannot be broken down or analyzed.

Mechanistic explanation, where it is applicable, is very satisfying. It is not surprising that people should try to extend it as much as possible, but it is not the only kind of explanation there is. Where it is inadequate, something has to give; we must either force things into explanations that can never fit them, or give up the comfort they might bring. At bottom, the post-Cartesian accounts of perception are no different in kind from the pre-Socratic ones that said the soul is made of all the material elements, so as to perceive them by registering like by like. These can be very pleasing and useful explanations of, at most, necessary underlying conditions of perception. Perceiving itself names a different act from chemical reaction or mechanical excitation, the being-at-work of a different sort of potency altogether. Aristotle finds this difference in part in the same double layering we meet in our explanations, since to perceive, a human or an animal must have bodily organs that take on the way of being-at-work of external perceptible bodies, but it must also have the nonbodily potency to be at work judging and discriminating those forms themselves (426b 8–14). This is a primordial kind of thinking, without which there is no perceiving in the first place. The crux of the matter, then, is here. Do we recognize this sort of rising above the behavior of mere bodies in animals? If so, in order to speak truly of animals, we must speak of them in the language of the activities of soul. Whatever perceives, Aristotle says, also imagines and desires, feels pleasure and pain, and can be spirited and have wishes (413b 22–24, 414b 1–3). Animals are our kin, and if we don't know them, we cannot know ourselves. Explanation by reduction cannot work here, but to see, with ever more clarity, how and why it fails, is also a work of explanation.

The two principal responsibilities that we noted as attributed to the soul are perception and the initiation of motion. We have seen that the structure of all life, plant and animal alike, the structure of something at work staying itself, applies to perception, and we have seen that working out this thought is itself a kind of explanation. For Aristotle, this explanatory structure applies even more directly to the initiation of motion. In Book VIII of the *Physics*, Aristotle concludes at the end of Chapter 5 that all motion must have its origin in something not in motion. His predecessors had explained the fact that animals are self-moving by conceiving the soul to be a body made of particles so small and smooth that they cannot come to rest, comparing them to the dust motes in a sunbeam. This is again a familiar and comfortable sort of explanation: one thing moves because some other thing is already moving and bumps it. If this leads to any paradoxes, we imagine them to be infinitely far back in time, where it is safe to ignore them.

Aristotle, though, is interested not in what came first in time, but in what stands first now and always as responsible for a motion, as governing its character and course. If motions happen reliably, in a pattern that recurs, there must be something holding on in a constant way that holds them in those patterns. So it is not only possible for a motion to be molded into a steady being-at-work, as in the case of a regular orbit, it is necessary for any and every motion that is identifiable to be traced back to something both active and constant. The very definition of motion (*Physics*, 201a 10–11) requires that there be potencies in things, and that in turn requires that those things be held together and maintained as beings of definite kinds, carrying those potencies in them.

It is clear that living things are just such beings, that actively maintain themselves according to kinds. And we have seen the sense in which animals, because they have perception, have within them a higher level of being-at-work that always implies some primitive kind of thinking. The structure of a perceiving being, then, is exactly what is required of a self-moving being, and Aristotle says exactly this in Book I of *On the Soul*, that the soul must move the animal by means of some sort of choosing and thinking (406b 24–25). He expands this thought in Book III, where he attributes the source of an animal's motion to the action of desire, but only when the desired thing is held sufficiently still in the animal's thinking or imagining for the desire to act at all (433b 10–13, 434a 14–16). Thus it is not merely a convenient and fortunate fact that perceiving and self-motion go together in animals, but a structure of activity that stands upon an intelligible structure of explanation. In the *Physics* (258a 1–5), Aristotle argues that any self-moving thing must have two parts, one of which (A) is in its own right unmoving but active, and responsible for the motion of the compound whole (AB). Putting together the arguments in the *Physics* and *On the Soul*, we can see that this amounts to a deduction of the soul. The necessary structure of any self-mover is AB, where B is an organized body capable of acting as a whole, and A is the higher being-at-work of the perceiving, and hence imagining and desiring, soul, which can initiate motion.

Again a mechanistic sort of explanation fails, and the recognition of that failure requires the discovery of a more adequate kind of explanation. One of Aristotle's objections to the claim that the soul is just a subtle body, already and always in motion, is that it cannot explain why the animal brings itself to rest (406b 22–24, *Physics* 255a 6–9). A billiard ball can make another billiard ball move, but it cannot make it stop at an intended place. That is just what self-motion means, though: the power to initiate a particular motion to an intended goal. And that is what we observe in animals. The subtle interior body in motion would give us something like a balloon that is blown up and released without being tied. It moves until it runs out of gas, in crazy gyrations wholly determined by the interior air, and for that very reason not determined by the moving thing itself, as a whole, in its own right. Many post-Cartesian philosophers decide that, in the world of bodies, all motion is completely determined for all eternity, so that our sense of being able to originate our own motion by choice must be illusory. Again, as with perception, the choice of a certain kind of explanation has led some people to declare that the thing that was to be explained is impossible. Immanuel Kant, to save the possibility of human freedom, locates its source in a supersensible world, since the world of experience must be an inflexible realm in which everything is determined. But the world I experience does not come laden with Cartesian and Kantian assumptions, and in it every bird and every ant refutes the claim of causal determinism by starting and stopping itself. The air and all the other fluids inside and outside an animal go merrily on with their motions according to their necessities, but as long as this does not amount to a hurricane or a flood, the animal remains in charge of where it goes. If this makes sense to you, Aristotle has given you a way to think about it.

From our vantage point in time, we can see that a way of looking at the world that turns over all explanation to the motions of bodies blinds itself to some of the most interesting possibilities of the world. Living comes about just where material bodies cease to explain anything, where they are organized into active wholes. It used to be said that the human body is mostly water while the rest of it is $1.98 worth of chemicals. Obviously, those materials, which could be collected in a bucket, are transformed when they are a human body, and only the form can explain the difference. That form is not a static arrangement but a being-at-work of organized material. Secondly, in an animal body, that first sort of being-at-work opens the way to a second, a receptivity to the world by which it encompasses other forms, and is at work discriminating them, as a form of forms, a perceiving being. Thirdly, that second kind of being-at-work makes it possible for the perceiving thing to fix its desires on objects, making it an independent and original source of motions that are not consequences of previous motions. This whole complex of interdependent ways of being at work opens out above the levels of material and motion, where material is molded into form and motion into active and energetic stillness. Without material and motion there is no life or soul, but without the nonmaterial and unmoving being-at-work of the soul there is no world at all that we would recognize.

The Ensouled Body

If an adequate account of the world must include souls, why is it that accounts that do include them seem even less believable than those that deny them? The *Timaeus* of Plato, for example, makes the soul a primary constituent of the whole of things, but seems addressed not to our belief, but rather to our pleasure in intelligent storytelling; the story is meant seriously but not literally, approaching deep questions in a playful spirit. In Plato's *Phaedrus,* Socrates says that he would rather hear myths about the soul than rational criticisms of those myths, because it is self-knowledge that is at stake (229B–230A). While accounts of one sort suffer from the deficiencies of reducing things to bodies in motion, those of the other sort seem to err in the opposite direction, by the excesses of fictional inventions. Aristotle says that the *Timaeus* involves an absurdity that afflicts most discourses on the soul; it speaks as though the soul in some way stands on its own and can be attached to or inserted in a body separate from and independent of it (407b 13–17), or even migrate from one body to another.

Much of Aristotle's thinking involves finding a path that steers between two alternatives that seem to exhaust the possible ways of understanding something. In Book XIII, Chapters 1–3, of the *Metaphysics,* Aristotle enters the dispute about whether mathematical things are separate from perceptible bodies or in them; he denies both claims. In the central argument of the *Metaphysics,* which runs through Books VII, VIII, and IX, he confronts the dilemma that the forms of things, since they are knowable, must be universal ideas, present only in thought, though at the same time they must be causes, acting independently of thought. In Book II, Chapter 3, of the *Physics,* he finds a definition of motion, despite the fact that the very nature of definition seems to fix things within static boundaries that prevent the possibility of motion. In each of these instances, Aristotle finds a way through that deepens the discussion of the topic, and in all of them some consideration involving potency and being-at-work allows him to resolve the impasse. The same thing is true of his way of getting hold of the soul.

It would seem that there are only two possibilities: either

there is nothing at all other than bodies, or there are distinct and detachable nonbodily entities that can be the souls of those bodies. The second half of this dichotomy has been ridiculed in this century as the idea of the ghost in the machine. Aristotle uses the gentler metaphor of the sailor in the boat. By sheer accident, most translations of *On the Soul* imply that Aristotle left open the possibility that the soul operates the body in the way that the sailor operates the boat. Aristotle's syntax at 413b 8–9 is sloppy, and allows a conclusive denial to be misread as a statement of uncertainty. But in its immediate context, his sentence picks up one that ends three lines above it. Together they say that, since the soul is the being-at-work of the body in the way that cutting is the being-at-work of the axe, or seeing is that of the eye, it is not hard to see why the soul is inseparable from the body, though if it were the being-at-work of the body in the way that the sailor is the being-at-work of the boat, that would be hard to see. The reference to the sailor is a mild joke, expressed in a contrary-to-fact conditional construction, the point of which is so obvious to its writer that he uses no verbs. The sailor is a metaphor for the detachable soul that Aristotle earlier called absurd (407b 13); the more accurate (though still inadequate) metaphor Aristotle himself has just offered is that of a design molded into wax (412b 7).

But while the separation of soul from body is absurd, the failure to distinguish them is equally mistaken. A living thing is a unity in the strongest possible sense of unity (412b 8–9), but its unity is made possible only by a certain sort of duality. We have seen above the argument from the *Physics* that a self-moving thing must have an AB structure, in which the whole moves in virtue of a part that is unmoving in its own right. And more generally, Aristotle says in the *Politics*, anything natural that is composed of parts, continuous or discrete, but that becomes one whole, has in it something that governs it and something else that is governed (1254a 28–32). But there is an inevitable ambiguity in speaking of the "part" that governs the whole or initiates its motion. In Book VII, Chapter 17 of the *Metaphysics,* Aristotle compares anything that is a genuine whole, and not a mere heap, to a syllable (1041b 11–33); just as it is not one of the letters that makes BA one syllable, it cannot be one of the elements or parts in any whole that unifies it. The thinghood of the whole thing takes precedence over all the parts, and is neither among them nor the mere sum of them. The whole appropriates and alters each of the parts, so that I cannot pronounce any letter without having in mind the syllable it must be shaped to fit. Only in the being-at-work of the whole is the part truly a part, so that the very nature of parts demands the presence of something that is not a part. In the syllable BA, the two letters are the parts which must have a third something uniting them, but in the self-moving body AB, B is the sum of parts that has the soul A uniting it.

Just as any sort of being-at-work is a stillness maintained in and through motion, any true wholeness is a unity present in multiplicity. The parts of the living body are many, but their life is one. The animal or plant is one as an independent thing and a *this* (412a 6–9), and its soul is the thing*hood* responsible for its *this*ness (412b 10–11). That is why, as soon as Aristotle has formulated a general definition of soul in the first chapter of Book II of *On the Soul,* he moves away from it toward the description of the various ways living things are at work. Life is manifest in the particularity to which the whole of *On the Soul* is only an introduction and a guide. It is in three massive works of observation, Aristotle's *History of Animals, Parts of Animals,* and *Generation of Animals,* together nearly ten times the length of *On the Soul,* that souls are on display, not as incorporeal fictions, but in all the vividness of embodied life. With the same passion with which Socrates loved the myths about the soul, and for the same reason, Aristotle was devoted to the fullness of knowledge of animals at work being themselves. In everything it does, a living thing presents its soul to view. The inseparability of body and soul is captured by the metaphor of a design pressed into wax, but their life is lost. In Book I, Chapter 9, of the *Physics,* Aristotle says that natural material is not something passive or resistant to its form, but that the material itself stretches out toward and yearns for form (192a 16–19). At the end of Book VIII of the *Metaphysics,* he says that the highest kind of material is one and the same thing as its form (1045b 17–19). In both places he is talking about living things as the primary examples of nature and of being. It is not just that in them body and soul are tightly bound into a whole, but that in them body extends to *mean* ensouled body, and soul draws body up to that fullness of meaning.

In Aristotle's view, inert, merely extended, body is a mathematical fiction, useful for limited purposes but always in need of correction when conclusions about the world are to be drawn. An example of the errors that follow from a persistent faith in inert matter may be found in the way the early development of living things is explained. Charles Darwin believed that some tiny bit of material had to pass over into the offspring from every cell of each of its parents, so that the eventual development of any structure or characteristic could result from the accretion of like material.[4] But he was one of the last to believe this, as the microscopic study of embryos confirmed that they develop out of material originally much less differentiated than it becomes later. But does this mean the embryonic material must be conceived as having some active power within it,

4. This hypothesis of "pangenesis" may be found in *The Descent of Man,* Modern Library Giant edition, p. 584. A fuller argument in favor of it, from an earlier work of Darwin's, is reproduced in *A Source Book in Animal Biology* (Cambridge: Harvard University Press, 1951), pp. 630–634.

capable of transforming it?[5] A new way of avoiding this conclusion was sketched out by the physicist Erwin Schrödinger in his 1944 book *What is Life?* (Cambridge University Press, 1985). If the chromosomes, of which every body cell has a complete set, are passive templates of every kind of material of which any part of the body will be made, all that is necessary is that each chromosome be a molecule that inertly keeps its order intact, or nearly so, through duplications and generations. But while a whole series of discoveries were making this hypothesis seem triumphant, the work of Barbara McClintock went unrecognized for several decades. She found that the genome actively rearranges itself, thus turning the materialist implications of DNA theory upside down. The chromosome is itself an organized body at-work, maintaining itself not as dead matter but with an active self-transforming power. The work of John Cairns, who since 1988 has repeatedly demonstrated that the genome can even direct its own mutation in the service of the organism, is now suffering the lack of acceptance of facts that don't fit the prevailing dogma.[6]

While Cartesian matter can contain only external differences, spatially in different places or temporally in succession,

---

5. Aristotle's classic argument to this conclusion, made without benefit of the microscope, is in his *Generation of Animals,* 733b 23–735a 29. It is restated in the language of cell biology by E. S. Russell in the book *The Interpretation of Development and Heredity* (Oxford, Clarendon Press, 1930, Chap. II), and vigorously defended by W. E. Ritter in the article "Why Aristotle Invented the Word Entelecheia," in The Quarterly Review of Biology (Vol. VII, No. 4, December, 1932, especially pages 388–393).

6. The original reports of McClintock's work with corn, and of that of Cairns with E. coli bacteria, may be too technical for the general reader, but two books by Evelyn Fox Keller contain helpful accounts. *A Feeling for the Organism* (W. H. Freeman & Co.,1983) is a biography of McClintock, and the last essay in *Secrets of Life, Secrets of Death* (Routledge, 1992) details the controversy Cairns ignited. A brief explanation of "Cairnsian mutation" appeared in the Washington Post, April 20, 1992, on page A3.

the ensouled body has internal differences in the same material at the same time, as potencies. The development of any embryo shows that these are not mere possibilities that might be brought about by external events, but inner natures that unfold out of causes within the organism. But at any time in the life of a mature organism there is a similar diversity of inner potencies manifest in health and disease. Countless advances in medicine have followed from the study of the body as a chemical and mechanical system; nearly every week brings us news that another disorder of our bodies or personalities has been unmasked as nothing more than a missing chemical or misfiring mechanism. The neurologist Oliver Sacks has been involved in the early application of more than one wonder drug that promised to overcome a severe disorder. But the material basis of Parkinson's disease or Tourette's syndrome is not the sole cause of the sufferer's trouble, and altering that material cannot be the sole cure. "Health and disease are alive and dynamic, with powers and propensities and 'wills' of their own," Sacks writes, and the return from severe disease requires understanding and strategies of accommodation. But physicians who have succumbed to the "madness" of the "Newtonian-Lockean-Cartesian view . . . which reduces men to machines, automata, puppets, dolls, blank tablets, formulae, ciphers, systems, and reflexes" become "overdeveloped in mechanical competence, but lacking in biological intelligence, intuition, awareness."[7] Altering the material condition of the

---

7. Oliver Sacks, *Awakenings* (Dutton Obelisk books, 1983), pp. 210, 205, 240–241, 247. See also the case history of "Witty Ticcy Ray" in Sacks's *The Man Who Mistook His Wife for a Hat* (Simon & Schuster Summit books, 1985). The diseases Sacks has worked with cannot be overcome without drugs, nor can they be overcome with drugs alone; only talk and understanding give the body a chance to find its way toward health. A rational medicine, he concludes, does not "attack" the disease in a *patient*, a passive medium, but collaborates with a sick human being who must be an agent of any cure.

body is only a beginning of treatment; if it is expected to be the whole of it, setbacks, frustration, and confusion are likely to follow. As experienced human beings, physicians may know that the healthy body is more than the sum of its material parts, but they must apply such knowledge outside the framework of any theory they have been taught. Studies occasionally report that such factors as having pets, having religious faith, and having a sense of humor contribute to the health of human beings, but conclusions such as these cannot be assimilated to the prevailing conception of body. The mathematical fiction of matter is so clear that it drives out the more difficult thought of the ensouled bodies that all our experience tells us we are.

Just because "biological intelligence" or "a feeling for the organism" saves the observer from underestimating the body, it makes the soul accessible to sight, and Aristotle was an observer of souls. There is a species of Mediterranean catfish described by Aristotle that now bears his name; his description was not accepted as fact until it was confirmed in the nineteenth century. Why? Because he reported that the male incubates the eggs and guards the hatchlings. The species is distinguished not by how it looks, but by how it lives, and for Aristotle, being-at-work is not a mere theoretical notion but an observable fact. Linda Wiener, an entomologist who finds that much of Aristotle's work as a biologist is at least the equal of anything done since, tells this story and the following one, similar to it.[8]

---

8. See "Of Lice and Men: Aristotle's Biological Treatises," in *The St. John's Review*, Vol. XL, No. 1 (1990–91), pp. 40–42. Biologists who have departed from many of Aristotle's particular teachings have still admired his fundamental understanding of living things. They include William Harvey, whose account of the circulation of the blood was adopted by Descartes as a model of mechanical explanation, and D'Arcy Thompson, who applied Newtonian physics to the living body, as well as Charles Darwin, whose assessment Wiener quotes (p. 39).

Not until the modern era did anyone repeat Aristotle's observation of the complex reproductive system of the smooth shark, which produces eggs, as fish do, but still bears its young alive, as mammals do. According to Adolf Portmann, Aristotle was the first writer to describe the fact that certain animals mark out, occupy, and defend definite territories, though the importance of the observation did not "register" with biologists until the twentieth century.[9] Even then it became interesting primarily as a species-preserving mechanism in natural selection.

It is always the case that, in the wide consensus of authoritative opinion, theory will take precedence over observed fact. A fact will be accepted only if it fits into some broader scheme of explanation. Aristotle's thinking about living things as ensouled beings permits him to see them as they are because he views each of them as an end in itself. Aristotle's teleology is often misunderstood. The *telos* for the sake of which each living thing does everything it does, on his view, is the wholeness of its own activity, complete in itself and not serving any external purpose. Aristotle is notorious for his alleged belief that all animals and plants are meant for use by human beings; he does say this, but only in the *Politics* (1256b 15–22), in a chapter that he says will follow "the beaten path" of popular opinion about property and ownership (1256a 2).[10] In the *Physics* he says more

---

9. See *Animals as Social Beings* (Viking, 1961), p. 25.

10. The phrase I have translated as "the beaten path" is usually translated as "our usual method"; versions of it are used by Aristotle only a handful of times. Near the beginning of the *Politics*, at 1252a 17–18 a similar phrase refers to an attempt to explain a whole by its parts, a decidedly *un*usual method for Aristotle, whose usual procedure is governed by the understanding that "the parts come from the whole" (*Metaphysics* 1023a 32). In the *Nicomachean Ethics*, the phrase (1108a 3) introduces a cursory and schematic arrangement of minor virtues and vices, of the "here the lecturer exhibits a diagram" sort. In the *Parts of Animals* (643b 11) and the *Economics* (1344a 10), it refers explicitly to being guided by widespread popular opinion.

clearly that human beings treat all other things as though they were for our sake (194a 34–36). There and in all his major theoretical works, Aristotle distinguishes between the end that is proper to a thing and the external end that it may also serve—another species, its own offspring (*On the Soul*, 415b 2–3), or the order of the cosmos (*Metaphysics* 1072b 1–3)—but his compact way of expressing this distinction, by the mere shift from a genitive to a dative pronoun, obscures for us what was plain as day to him. F. H. A. Marshall, in the foreword to the Loeb Library edition of the *Parts of Animals* (in reference to 696b 24–35), says that only once in all his writings (in considering the position of the mouth in sharks) does Aristotle even tentatively and partially ascribe to any organic structure a function external to its own species. Living things do not *have* purposes, they *are* purposes.

In Aristotle's teleology, every part of the living body has a function, to maintain that body and its activity, and much of that activity serves as means to some end other than itself, but finally there is some sort of activity that has no function. Bulldogs, for example, were bred for the purpose of fighting bulls, but according to Vicki Hearne, what bulldogs of any breed do out of their own strongest desire is climb on top of anything, from an overturned washtub to a mountain, that can be surmounted.[11] Adolf Portmann devotes the last chapter of *Animals as Social Beings* to the Australian lyre-bird, a species

11. See *Bandit, Dossier of a Dangerous Dog* (Harper Collins, 1991), pp. 58–68. It is especially instructive to learn that the artificial enhancement of certain traits by the breeder may result in the emergence of an altogether different natural trait; this is natural selection in a sense not intended by Darwin. Darwin cites as typical the attitude of the pigeon breeder Sir John Sebright, that "he would produce any given feather in three years, but it would take him six years to obtain head and beak." This is quoted in the first edition of *The Origin of Species* (in Penguin Classics edition, 1987, p. 90), but later removed. One suspects that the experience in the 1920s of the Russian geneticist Karpechenko may be the more usual result; he crossed a radish

that has a rich family life in which the female does most of the practical work, while the male contributes by singing and dancing, through a varied repertoire, on platforms constructed as carefully as are the nests, for months out of the year. These performances are not to attract a mate or compete with a sexual rival, and while they may be territorial in origin, Portmann interprets them as expressive of the emotional life of the whole family. Like Aristotle, Portmann in all his work has an eye for those manifestations of life that are for their own sake, in which some living thing is simply and purely at work being itself.

The Thinking Soul

To speak of an activity as being for its own sake means that the one engaged in it feels complete or fulfilled when at work in that way. The ability to come to rest in such an activity is the counterpart of the ability to initiate one's own motion, and is impossible without some sort of thinking and choosing, as Aristotle says (406b 22–25), but in the other animals this may be understood as guided entirely by a sensory imagination (434a 5–7). Vicki Hearne, in the passage cited above, speaks of the bulldog in the presence of a mountain or boulder as feeling the call of the sacred or the sublime. Indeed, it seems that imperatives, or invitations, of such a kind could make themselves known to us in no other way. The difference of the human animal would not be in the manner in which the desired state calls to us, but in the content of it and in our way of responding to it. Properly human work could not leave our distinctive capacities unfulfilled. In Book I of the *Nicomachean Ethics*, Aristotle's first outline sketch of the human end says

with a cabbage and obtained a vigorous plant entirely inedible by humans, with the root of a cabbage and the upper parts of a radish.

that it must be some way of putting into action our rationality (1098a 3–4).

The characteristic human capacity is manifest as speech, which is not the same thing as communication. The other animals perceive things as pleasant and painful and communicate those perceptions to one another, but lack the capacity to distinguish among those pleasant and painful things as advantageous or disadvantageous (*Politics* 1253a 7–18). To do the latter would require combining the images in the imagination into wholes that can be compared to a single criterion of the greater good (*On the Soul* 434a 7–10). This would amount to holding on to the universal kinds within particular perceptions (*Posterior Analytics* 100a 5–9), and connecting these in judgments or propositions (*On Interpretation* 16b 26–17a 8). Such a judgment, the elementary act of reasoning, is embodied in a sentence, the minimal unit of speech. Speech, then, in Greek *logos,* is the same capacity as *to logistikon,* the faculty of reasoning, or reckoning things together, and is the medium of discursive thinking—*dianoia*—or thinking things through. This activity of thinking opens a wide possibility of error, since it is not a mere bodily process, but has a content, an understanding that is inherently right or wrong (*On the Soul* 427a 17–b 11). This new field for error is also a new realm of freedom, both in pursuing practical ends and in rising above practical pursuits altogether to embrace knowledge as an end for its own sake (*Metaphysics,* Book I, Chapter 1).

The ways in which the thinking embodied in speech can be rightly or mistakenly used are explored by Aristotle in a collection of six works known traditionally as the *Organon* (*Categories, On Interpretation, Prior* and *Posterior Analytics, Topics,* and *On Sophistical Refutations*), a name that means instrument, and separates these logical writings as subordinate to the proper work of philosophic inquiry. The experiences of the soul on which speech is founded, Aristotle says at the beginning of *On Interpretation* (16a 8–9), belong to a different study, treated in *On the Soul*. But the whole topic of thinking things through is scarcely mentioned at all in *On the Soul*, though it permeates the life of the human soul. Moreover, the way in which all the powers of the human soul can be developed and combined is addressed not in *On the Soul* but in the *Nicomachean Ethics*. The whole consideration of thinking in *On the Soul* revolves around one question alone: what is the relation between the thinking soul and other beings that makes thinking possible? This is not a "psychological" question in any sense in which that word is now used. How could any evidence to answer it be found by any sort of introspective, experimental, or clinical observation? The interpretation of any such observation would presuppose assumptions about the kinds of beings that make up the world and the kinds of relations among them that are and are not possible. These are ultimate topics of philosophic reflection. Their presence in *On the Soul* lifts that work briefly into what Aristotle calls first philosophy, and leads to its most memorable, and most difficult, assertions.

From the beginning of *On the Soul*, Aristotle prepares the way for the difficulties that underlie an adequate account of thinking. In Book I, Chapter 1, he seems to anticipate including thinking among the activities of the ensouled body. He says that, while thinking might seem to be an exceptional activity, belonging to the soul alone, it must still involve the body if it always involves imagination (403a 8–10). But in Chapter 4 of Book I, he says that the intellect by which we think seems to be an independent thing within us, indestructible and unaffected by what happens to the body (408b 18–19, 29). Since he eventually concludes that all thinking is connected with imagination (431a 16–17; 432a 8–9, 13–14), it might seem he would have to decide that it is inseparable from the body.[12] But even when he declares the unity of soul and body most emphatically, likening it to that of an eye and its seeing or an axe and its cutting, he reserves the possibility that something in the soul might be altogether independent of the body (413a 6–7). And finally he concludes explicitly, in Book III, Chapter 5, that there is a separate, deathless, everlasting intellect.

What is this intellect that is so difficult to locate? It is that which thinks things as indivisible and whole, as distinct from thinking them step-by-step in time (430a 26), so that its thinking is more like rest than motion (407a 32–33). This contemplative thinking (*noēsis*) of the intellect (*nous*) thus stands opposed to thinking things through (*dianoia*), but it also stands beneath the act of thinking things through and makes it possible. Every judgment is an external combination of a separated subject and predicate in our discursive thinking, but is simultaneously held together as a unity by the intellect (430b 5–6). That is why Aristotle says that the contemplative intellect is that by means of which the soul thinks things through and understands (429a 23). Its thinking is the foundation upon which all other thinking proceeds, just as having our feet on the ground is one of the conditions of our walking. Exclusively discursive thinking that could separate and combine, but could never contemplate anything whole, would be an empty algebra, a formalism that could not be applied to anything. In human thinking, at any rate, the activities of reasoning and contemplation are rarely disentangled.

This twofoldness within the meaning of thinking causes an

12. In *On the Soul*, the connection of thinking with imagination is asserted in limited contexts, which one must explore to see what that connection is, but the unqualified claim that there is no thinking without an image is made in *On Memory and Recollection*, at 449b 31.

unavoidable confusion, which can be compounded by translations that are not attentive to the difficulty. In particular, the use of the word "mind" can muddle things beyond repair. The idea of mind is an orphan left behind by the Cartesian shift in the conception of body. The mind is a kind of container, isolated and self-enclosed, in which thoughts, desires, and feelings—all the remnants unconnected with extended body—are found.[13] The closest thing to it in classical philosophy is the birdcage in Plato's *Theaetetus* (197A–200D), but that image is a parody meant to expose the absurdity of considering knowledge a stock of possessions rather than a living activity. As is often the case with topics of Aristotle's inquiries, the word *nous* and its verb *noein* have broad meanings at the beginning of *On the Soul* and narrower meanings as the work proceeds, but never does Aristotle construe the noun or verb as naming anything but an activity. As we shall see, even when Aristotle speaks of the intellect as passive, indeed as pure and unmixed passivity, he is still speaking of a high level of concentrated activity, in no way compatible with any notion of a mind stored with ideas.

The confusion that is not a product of translation has to do with whether there is an intellect at the foundation of the world, as well as an intellect that is part of the soul. Three times Aristotle refers to "what is called intellect." Once this is in relation to the assertion in Plato's *Timaeus* that there is a soul of the world; Aristotle claims that the functions given there to a world-soul really belong to what is called intellect (407a 4). A second time, it is connected with Anaxagoras's positing of an intellect that rules all things; in this case Aristotle adopts the word and the argument made about it, but applies them to something within our souls (429a 22). The third time, the phrase is equated with the reasoning part of the soul

13. See above, page 5.

(432b 26). The twofoldness described above, of discursive and contemplative thinking, is, for Aristotle, inseparable from the distinction between a human intellect and an intellect on which the world as a whole is founded. Aristotle never doubts that the discursive intellect that thinks things through belongs to the embodied soul, and decays and dies along with it (408b 25–29). But the discursive intellect is bound up with a contemplative intellect that must be in us but not of us. If it were not somehow in us, our thinking would not be what it is; if it were wholly within us and subject to our limitations, no thinking would be possible at all. This is the claim made in Book III, Chapter 5.

The productive or formative intellect spoken of in III, 5 has attributes that can belong only to an everlasting, unfailingly active being that is separate from all embodied life. But things are also said of it that can refer only to an activity engaged in by human beings. This is a paradoxical situation, but it is the other side of the coin to another paradox discovered elsewhere. The paradox in this chapter of *On the Soul* is hard to grasp because it is set out in so brief and sketchy a way. The other paradox that is its twin is presented in the *Metaphysics,* and is hard to grasp because it requires putting together all the parts of that sprawling work into one whole argument. The separate intellect uncovered in Book XII, Chapter 7, of the *Metaphysics* as the unchanging first mover of the cosmos is also the source of unity at work in every genuine being within the cosmos, that is, in every animal and plant, a source that is narrowed down through Books VII–IX, the argument of which is linked to Book XII by a crucial step in Book X (1052a 29–33). Thus every living thing is continuously a product of a divine intellect at work within it, while that same intellect is not reciprocally affected by it, but remains wholly separate and independent. The active intellect posited in *On the Soul* is this same source, looked at from the side of the thinking animal, the human being that is not only its product but can participate in its activity, by occasionally merging into the contemplation which is constant in it, yet still without affecting that constant activity. The *Metaphysics* and *On the Soul* are independent inquiries, neither of which presupposes the other, but they cohere, and interpretations of them confirm each other.[14] That the intellect is first of all and primarily a divine activity, but secondarily present in human beings, is reflected in Aristotle's claim in the *Generation of Animals* that, in the human embryo, everything belonging to the soul develops along with the body except the intellect, which comes in from outside (736b 27–29).

In the *Metaphysics*, the argument leading to the divine intellect has two dialectical beginnings, one inquiring into the way in which motion can have an unmoving source, the other asking how something compound can be a single whole. In *On the Soul*, the active and productive intellect is posited without argument, but that positing is surrounded by a group of assertions that set out the conditions of the possibility of contemplative thinking. These assertions seem best understood as postulates. They are not arrived at in Aristotle's usual way, by examining and refining the opinions of previous thinkers and widespread beliefs. In his inquiry about the soul, that sort of dialectic comes to an end early in Book III, Chapter 3, with the observation that all the ancients failed to consider how thinking differs from perceiving, assuming that it too is a bodily process and not subject to error. From that point on, Aristotle is in

---

14. The attempt by some scholars to arrange Aristotle's works in an order of "development" of his thought is always misguided. A remark in the *Metaphysics* at 1026a 4–6, implying that some part of the soul is not embodied, seems to depend on conclusions reached in *On the Soul*, but a remark in *On the Soul* at 408b 29, attributing divinity to the intellect, seems equally to depend upon conclusions of the *Metaphysics*. It is safest to assume that all Aristotle's major writings are products of his whole life of teaching.

uncharted territory.[15] William O'Grady has pointed out the exceptional character of Aristotle's writing about knowing,[16] its unusual tentativeness, its reliance on a certain kind of negative determination, and the strikingly frequent appearance in it of the adverb *pōs*—"somehow." O'Grady identifies five "vast and simple propositions" that Aristotle ventures about knowing:

1. The soul is none of the beings.
2. The soul is in a way all things.
3. The soul becomes identical with what it knows.
4. To know is somehow to be acted upon.
5. The full being-at-work of what acts resides in what is acted upon.

If these postulates correctly identify what must be explained to account for a being that is capable of contemplation, but is not always actively contemplating, there is need to acknowledge a higher source responsible for the original holding together of what is knowable, capable of joining with and putting to work the human potency. In this way the fact of human contemplation becomes a dialectical beginning for the discovery of the divine intellect, independent of the arguments in the *Metaphysics*. But it is important to note that it is not only the fully realized act of contemplation that establishes it as one of the powers present in human beings. As we have already argued,

---

15. In Plato's *Theaetetus*, the explicit question is "what is knowledge?", but, as Jacob Klein has shown, what is discussed there is in fact error. Klein's interpretation of the *Theaetetus* appeared in his book *A Commentary on Plato's Meno* (University of North Carolina Press, 1965, pp. 157–166), and later, expanded, in *Plato's Trilogy* (University of Chicago Press, 1977, pp. 75–145). The possibility of contemplative knowing lurks everywhere around the edges of that dialogue, which invites the reader to begin a process of questioning that Socrates fails to stimulate in Theaetetus or Theodorus.

16. This remarkable reading of the most difficult part of a difficult text is in the lecture "About Human Knowing," included with the collected writings of William O'Grady in The St. John's Review, Winter, 1986.

all discursive reasoning rests on a foundation of contemplative thinking, but that same foundation must also underlie human sense perception. In the *Nicomachean Ethics*, Aristotle says that the same capacity of intellect grasps both the primary intelligible things and the ultimate perceptible particulars (1143a 35–b 6). Again and again in that work, Aristotle speaks of the recognition of the right thing to do in particular circumstances as an act of perception (e.g., 1109b 23–24), and he likens this sort of perceiving to our immediate recognition that a triangle is the last kind of figure into which a polygon can be divided (1142a 28–30). In *On the Soul*, he calls this incidental perception, by which we recognize something white as the son of Diares (418a 20–21) or the son of Cleon (425a 25–27); his examples are not mere proper names, but perceptible wholes understood through the intelligible relation of a male offspring to its father. By calling this a kind of perception, even though it is incidental to the proper content of perceiving, Aristotle indicates that we do not construe the white blob by a series of steps, but take it in at once as what it is (even if we *mis*take it). This is an act of intellect, embedded in our simplest and humblest sensory experience.

Like our highest knowing, our perceiving takes in something organized and intelligible all at once and whole. That is why we can contemplate a scene or sight before us as well as something purely thinkable. In neither case is the thing grasped our product, and that is why Aristotle calls both perception and contemplative thinking passive (*pathētikos*), but this receptiveness to being acted upon should not be confused with inertness. It is rather an effortful holding of oneself in readiness. Attentive seeing or concentration in thinking requires work to keep oneself from distraction; it is a potent passivity (*dunamis*) that becomes activity in the presence of those things that feed it. Nutrition is the active transformation of things in the world into the living body; contemplation is the effortful openness of the soul to a merging into the intelligible foundation of the world. Reading and listening are always hard work, and hardest of all when one lets the meaning of the speaker or author develop within oneself. Contemplation, as Aristotle intends it, is the same sort of act without any building up of interpretation; it is rather what he calls affirming something not by thinking any proposition about it but by touching it (*Metaphysics* 1051b 24–25). Like the good reader, the contemplative knower is most active in one way (*energetikos*) just by being least active in another (*poiētikos*). This achievement of the inactivity of letting things be, joined with the activity of complete absorption in knowing things as they are, is definitely regarded by Aristotle as a human possibility (*Metaphysics* 1072b 14–15, 24–25; *Nicomachean Ethics* 1177b 34–1178a 2), though it must be small in time and "bulk" in relation to the whole of life. But even when we are not letting things be and are groping our way through error and confusion, Aristotle believes that the activity of knowing is always at work in us and available to us (*Physics* 247b 1–18), potentially guiding everything we do in the same way the blind grub worm is led to its food and a plant is turned to the light.

Notes on this Volume

Like my other translations of Aristotle,[17] this one avoids as much as possible the Latin tradition out of which most English versions of Aristotle have come, and seeks to recapture the vividness of the Greek original. Latinate translations of *On the Soul* suffer most from an inadequacy at the heart of the definition of soul, which, according to Aristotle, is an *entelecheia*,

17. See Aristotle's *Physics* (Rutgers University Press, 1995, 2nd paperback edition 1998) and Aristotle's *Metaphysics* (Green Lion Press, 1999).

a being-at-work-staying-itself. When this is translated as "actuality," the living meaning of the whole book is given a mortal blow. And since form, *eidos* or *morphē*, is understood by Aristotle as being-at-work (*energeia*), the interpretation of a central claim of the book, that sense perception is the reception of form without material, is deprived of its principal source. These things are discussed in the first part of this introduction, "Recognizing the Soul." Also, a whole array of misleading modern connotations is dragged into Aristotle's word *nous* (intellect or contemplative intellect) when it is translated as "mind." This is discussed in the third part of the introduction, "The Thinking Soul."

The oldest existing manuscripts of Aristotle's writings date back to about a thousand years ago. Those of *On the Soul* differ from one another to a much greater extent than do those of most of Aristotle's other works. This translation follows the edition of Sir David Ross (Aristotle, *De Anima,* Oxford, Clarendon Press, 1961) with only a few departures. The earlier edition of R. D. Hicks (*Aristoteles, De Anima,* 1907, reprinted by Georg Olms Verlag, Hildesheim, 1990) contains a wonderful treasure house of helpful notes, including quotations in Greek from the ancient commentators. The Loeb Library version of *On the Soul* contains and is translated from the earliest modern edition of the Greek text, published by Immanuel Bekker in 1831. (The page numbers and column letters from Bekker's works of Aristotle became the standard pagination carried now by all versions, and printed here in the margins; the line numbers with them correlate approximately with those in Ross's text.) Despite the great variability in the manuscript evidence, a comparison of versions based on different Greek editions will show few discrepancies of any great consequence; the most noticeable one is a confusion about imagination in bees, ants, and grub worms. Important matters of interpretation depend more on a sense for the whole and the convergence of many sources of evidence. To read Aristotle directly in any Greek edition is worthwhile; to consult the critical apparatus (variant readings at the bottom of each page) contained in some of them when a passage doesn't feel quite right may stimulate a new thought or resolve an occasional difficulty.

The small treatise *On Memory and Recollection* is one of a group of writings by Aristotle concerning various activities that belong to the soul and body together. It is included here as a valuable supplement to *On the Soul,* and it too is translated from an edition by Sir David Ross (Aristotle, *Parva Naturalia,* Oxford, Clarendon Press, 1955), again with some departures. For a discussion of the light this "wonderful little work" sheds on Aristotle's understanding of imagination, see Eva Brann's book *The World of the Imagination, Sum and Substance* (Rowman & Littlefield, 1991), pp. 43–44. For a discussion of the way it may be read as a "sober commentary" on the nonmythical content of Plato's *Meno,* see Jacob Klein's book *A Commentary on Plato's Meno* (University of North Carolina Press, 1965), pp. 109–112.

Rough descriptions of the contents of each chapter are listed in the table of contents, partial glossaries are given in notes before each of the three books of *On the Soul,* and a bibliographic note and a selective index may be found at the back of this book. For the convenience of the reader, a general Aristotelian glossary that applies to the *Physics* and *Metaphysics* as well as to *On the Soul* is also included.

Acknowledgments

I first experienced the delight of following Aristotle beyond the artificial boundaries of contemporary opinions in seminars thirty-five years ago led by Robert Bart. I continue to make discoveries in those regions all the time, and the delight has not faded. Formative guidance in my studies came first and best from a lecture by Jacob Klein, and later from a book by Joseph

Owens and transcribed lecture courses of Martin Heidegger, all cited in the bibliographic note below. Heidegger attempted to dismantle the whole philosophic tradition, and for that purpose sought to bring to life the insights at and near its beginning, including those of Aristotle. Klein was not a disciple of Heidegger, but was a student of his in a truer sense; he said much later that it was in hearing Heidegger lecture on Aristotle that he first discovered the possibility of understanding the thinking of another person. Owens writes from within the Latin Aristotelian tradition, with sufficient knowledge to uncover its errors, and with the intellectual integrity to be a greater friend to truth.

My own ventures into translation have been an effort to put the things I have learned from these teachers, and many others, directly into the English text. My classroom experience with other translations had been filled with the need to say "well, that word doesn't really mean that, but something more like . . . ." I began to wonder whether it wouldn't be possible for the written text to begin nearer to the living text that classroom discussion usually produced, through our artless, but serious and honest, attempts to be simply accurate. Thanks to repeated and generous support from St. John's College during a decade now, three translations have seen the light. It was the stern urging of Brother Robert Smith that led me to add *On the Soul* to the earlier translations of the *Physics* and *Metaphysics*. These are the three principal theoretical works of Aristotle, and of them *On the Soul* is the most challenging for the translator. In my first version of it, I succumbed more than once to the temptation to import more into the text of a crucial sentence than Aristotle had put there. Steven Werlin, who used that version in a class at Shimer College, made several suggestions that have improved this book, and helped me keep the line between text and footnotes in place. A number of the books and articles I refer to throughout the introduction and notes were made known to me by Kathleen Blits. My thanks go also to those invisible correspondents on the Aristotle e-mail discussion group (aristotle@lists.enteract.com) who helped me make the argument that *On Memory and Recollection* belongs in this book. I am grateful to many others, more than I can mention here, and especially to Dana Densmore and Bill Donahue for the thoughtful and painstaking work that has made the books from the Green Lion Press everything a serious reader could desire.

Annapolis, Maryland
Spring 2001

On the Soul

Introductory Note to Book I

Aristotle's first approach to his inquiry about the soul is to ask about its *thinghood*. This question involves a fundamental distinction among the ways *being* is meant, since anything that *is* must be either an *independent thing* or in some way dependent upon such a thing, as an *attribute*, a part, an *activity*, or a *potency* of it. An independent thing is a distinct item of experience, that presents itself to us as a *this*, intact and *separate* from its surroundings, and maintains itself as such through change. Thus, while the inquiry about the soul belongs primarily to the study of nature, the realm of things that move and change, its beginnings belong to *first philosophy*, the study of the unchanging constitution of the world and the beings within it. These kinds of study are distinguished by Aristotle in Book VI, Chapter 1 of the *Metaphysics*, and the topic of thinghood is introduced in Book VII, Chapter 1 of that work. Hence, the following vocabulary underlies the beginning of Book I of *On the Soul*:

τὸ ὄν (*to on*) being, that which is

οὐσία (*ousia*) thinghood, an independent thing

πάθος (*pathos*) attribute

ἔργον (*ergon*) work, activity

δύναμις (*dunamis*) potency

τόδε τι (*tode ti*) a *this*

χωριστόν (*chōriston*) separate

πρώτη φιλοσοφία (*prōtē philosophia*) first philosophy, metaphysics

Aristotle's way into any inquiry is to look for the *impasses* involved in the opinions about it held by ordinary people and by earlier thinkers. These are not mere "difficulties" or "perplexities" but stalemates in one's thinking that reveal the inadequacy of common presuppositions, and force one to a new and deeper way of looking at things. Most of the accounts of the soul given by earlier thinkers center around various conceptions of things as composed of *elements;* these are regarded by some as the smallest bodies—invisible and rapidly moving atoms—by others as the simplest kinds of bodies—earth, air, fire, and water—and by still others as the primary mathematical constituents of things as having number and magnitude in them. (Aristotle's own understanding of elements differs from all of these; see the note to 423b 30.) These elements by themselves are made responsible for both motion and perception, though some thinkers posit as an additional source a divine or cosmic *intellect,* while others speak of the soul as a *harmony* of bodily elements. The bulk of Book I thus involves discussion of the following:

ἀπορίαι (*aporiai*) impasses

στοιχεῖα (*stoicheia*) things in a row, letters, elements

νοῦς (*nous*) intellect

ἁρμονία (*harmonia*) a joining (in carpentry), an attunement, a harmony

Book I

Chapter 1 Since we consider knowledge to be something beautiful and honored, and one sort more so than another either on account of its precision or because it is about better and more wondrous things, on both these accounts we should with good reason rank the inquiry about the soul among the primary studies. And it seems that acquaintance with it contributes greatly toward all truth and especially toward the truth about nature, since the soul is in some way the governing source of living things. And we are seeking to bring to sight and to understanding the nature and thinghood of the soul, and then whatever follows about it, among which seem to be some attributes of the soul by itself, and others that belong to the living things, on account of the soul. 402a 402a 10

But altogether in every way the soul is one of the most difficult things to get any assurance about. For, since what is being sought is common to many other studies—I mean an account of the thinghood of something and of what it is—someone might perhaps think that there is some one method that applies to everything we want to discover the thinghood of, just as logical demonstration is the method of discovering the attributes peculiar to each thing as consequences of its thinghood. In that case, what would have to be sought would be this method, but if there is not some one common method of pursuing what something is, the work we have taken in hand becomes still more difficult, since it will require that we get hold of some way of approaching each particular thing. And even if it were evident whether this were demonstration or division [of general classes] or some other method, there would still 402a 20 be many impasses and false starts over the things from which one ought to begin the search, for different studies have different starting points, just as do the studies of numbers and of plane figures.

But first, perhaps, it is necessary to decide in which general class it is, and what it is—I mean whether it is an independent thing and a *this*, or a quality or quantity or some other one of the distinct ways of attributing being to anything, and further whether it belongs among things having being in potency or is rather some sort of being-at-work-staying-itself; for this makes no small difference. And one must also examine whether it is divisible or without parts, and whether all soul is of the same kind, or, if it is not of the same kind, whether souls differ as forms of one general class, or in their general classes. For those who now speak and inquire about the soul seem to consider only the human soul, but one must be on the lookout so that it does not escape notice whether there is one articulation of soul, just as of *living thing*, or a different one for each, as for horse, dog, human being, and god, while a living thing in general is either nothing at all or a later concern—as would similarly be in question if any other common name were applied.

402b

Again, if there are not many souls but parts of one soul, one must consider whether one ought to inquire first about the soul as a whole or about the parts. But it is difficult even to distinguish, among these, which sorts are by nature different from one another, and whether one ought first to inquire about the parts or about the work they do: the thinking or the intellect, the sensing or the sense, and so on in the other cases. But if the work the parts do comes first, one might next be at a loss whether one ought then to inquire about the objects of

402b 10

these, such as the thing sensed before the sense, and the thing thought before the intellect. But not only does it seem that knowing what something is would be useful for studying the causes of the things that follow from its thinghood (just as, in mathematics, it is useful to know what straight and curved are and what a line and a plane are, for learning how many right angles the angles of a triangle are equal to), but it seems too, on the contrary, that those properties that follow contribute in great part to knowing what the thing is, for it is when we are able to give an account of what is evident about the properties, either all or most of them, that we will also be able to speak most aptly about the thinghood of the thing. For in every demonstration the starting point is what something is, so it is clear that those definitions that do not lead to knowing the properties, nor even to making them easy to guess at, are formulated in a merely logical way and are all empty. [402b 20] [403a]

And there is also an impasse about the attributes of the soul, whether all of them belong in common to it and to the thing that *has* the soul, or any of them belong to the soul itself alone. It is necessary to take this up, though it is not easy, but it does seem that with most of its attributes, the soul neither does anything nor has anything done to it without the body, as with being angry, being confident, desiring, and every sort of sensing, though thinking seems most of all to belong to the soul by itself; but if this is also some sort of imagination, or cannot be without imagination, it would not be possible for even this to be without the body. Now if any of the kinds of work the soul does or any of the things that happen to it belong to it alone, it would be possible for the soul to be separated; but if nothing belongs to it alone, it could not be separate, [403a 10] but in the same way that many things are properties of the straight line *as* straight, such as touching a sphere at a point, still no separated straight line will touch a bronze sphere in that way, since it is inseparable, if it is always with some sort of body.

But all the attributes of the soul seem also to be with a body—spiritedness, gentleness, fear, pity, boldness, and also joy, as well as loving and hating—for together with these the body undergoes something. This is re-
403a 20
vealed when strong and obvious experiences do not lead to the soul's being provoked or frightened, while sometimes it is moved by small and obscure ones, when the body is in an excited state and bears itself in the way it does when it is angry. And this makes it still more clear: for when *nothing* frightening is happening there arise among the feelings of the soul those of one who is frightened. But if this is so, it is evident that the attributes of the soul have materiality in the very statements of them, so that their definitions would be of this sort: being angry is a certain motion of such-and-such a body or part or faculty, moved by this for the sake of that. So already on this account the study concerning the soul belongs to the one who studies nature, either all soul or at least this sort of soul.

But the one who studies nature and the logician
403a 30
would define each attribute of the soul differently, for instance what anger is. The one would say it is a craving for revenge, or some such thing, while the other would
403b
say it is a boiling of the blood and heat around the heart. Of these, the one gives an account of the material, the other of the form and meaning. For the one is the articulation of the thing, but this has to be in a certain sort of material if it is to be at all. In the same way, while the meaning of a house is of this sort, a shelter

that protects from damage by wind, rain, and the sun's heat, another person will say that it is stones, bricks, and lumber, and yet another that the form is in these latter things for the sake of those former ones.

Which of these is the one who studies nature? Is it the one concerned with the material who ignores the meaning or the one concerned with the meaning alone? Or is it rather the one who is concerned with what arises out of both? What, then, are each of the others? Or is there not just one sort of person concerned with the attributes of material that are not separate nor even treated as separate, but the one who studies nature is concerned with *all* the work done by and things done to a certain kind of body or material; but someone else is concerned with whatever is not of that sort, with certain of them a skilled worker if it happens to be the concern of, say, a carpenter or doctor. But the mathematician is concerned with things that are not separate, but not *as* attributes of a certain sort of body but as taken away from bodies, while the one who studies first philosophy[1] is concerned with things insofar as they *are* separated. But we were saying that the attributes of the soul are inseparable from the natural material of living things in the former way, at least those that are present in such a way as spiritedness and fear, and not in the way a line or a plane are inseparable. 403b 10

Chapter 2 While inquiring about the soul, we must bring up the impasses that we need to find a way past in order to move forward, and at the same time take up 403b 20

1. This is what we, following the usage of ancient librarians, call metaphysics.

along the way the opinions of all those earlier people who have made any claims about the soul, so that we might take hold of the things that have been said rightly while, if any are not right, being wary of these. And the start of the inquiry is to set forth the things that most of all seem to belong by nature to the soul. Now that which is ensouled seems to differ from that which is without a soul in two ways most of all: in motion and in sensing. And these two things are just about what we have received from our predecessors concerning the soul. For some say that, first of all and most of all, the soul is that which produces motion, and since 403b 30 they supposed that what was not itself in motion could not move something else, they assumed the soul to be something among the things that are in motion.

404a Hence Democritus says that it is some sort of fire or heat, for of the infinitely many shapes, or atoms, there are, he calls those that are of a spherical form fire or soul (like the so-called motes of dust in the air, which appear in the sunbeams that come in through windows); and he says that all such seeds together are the elements of the whole of nature (as similarly does Leucippus), and that of these the spherical ones are the soul since such shapes most of all are able to slip between everything else and to move the rest, while moving also themselves, since they suppose that the soul is what provides motion to animals. This is also why they suppose that breathing is the distinctive mark 404a 10 of life, for when the surrounding bodies compress the animal and squeeze out the atomic shapes that provide it with motion—by not themselves ever being at rest—help comes from outside when other such shapes enter in turn in the act of breathing, for these also prevent the ones that are already in the animals from being emitted, by holding back the compression and compacting, and it lives as long as it is able to do this.

And what is said by the Pythagoreans seems to have the same thinking in it, for some of them said that the soul is the specks in the air, while others said it is what moves them, and about these they said that these specks are manifestly in motion continuously, even when the air is completely still. And it comes to the same thing with all those who say that the soul is what moves itself, for they all seem to assume that motion is the thing that is most proper to the soul, and that, while all other things are moved by the soul, this is moved by itself, because they do not see anything cause motion which is not also itself in motion. In a similar way, Anaxagoras too says that the soul is the thing that causes motion, as does anyone else who has said that the totality of things was set in motion by intellect, though certainly not altogether in the way Democritus meant it, for *he* took soul and intellect to be simply the same thing. (For he says that the appearance is the truth, and that for this reason Homer rightly made Hector "lie thinking changed thoughts."[2] So he does not use "intellect" as any sort of potency concerning truth, but says that the soul and the intellect are the same thing.) Anaxagoras is less clear about them, for in many places he says that the intellect is responsible for what is beautiful and right, but elsewhere he says that the intellect is the same as the soul, since he says that it is present in all animals, the great and the small, of high or low rank; but there is no evidence that intellect, if meant at any rate in the

404a 20

404a 30

404b

2. In fact it is Euryalus, knocked out by Epeius during the games, whose delirium is described this way at *Iliad* XXIII, 698.

sense of thoughtfulness, is present in all animals alike, nor even in all human beings.

Those, then, who had their eyes on the fact that the ensouled thing is in motion, all supposed that the soul is that which is most suited to produce motion, but those who look to the knowing and perceiving of beings

404b 10

all say that the soul is the ultimate sources of things, some making them more than one, others making it one. In this way Empedocles made the soul out of all the elements, but also said that each of these is soul, saying

> We behold earth by means of earth, water by means of water,
> Ether by bright ether, but fire by sight-destroying fire,
> Love by means of love, and strife by ruinous strife.

And in the same way, Plato in the *Timaeus* makes the soul out of the elements, since like is known by like and the things are made of the ultimate sources. Similarly

404b 20

also in the writings on philosophy[3] it is spelled out that animal itself is composed of the form itself of the one, and of the primal length, breadth, and depth, and other things are composed in a similar way; in another way as well, intellect is the one and knowledge the dyad (since it is one directional toward one end), while opinion is the number of the plane, and sense perception of the solid. For the forms themselves and the ultimate sources

---

3. There are various guesses about this source, which is not extant; it is usually taken to be by Aristotle rather than Plato. More extensive discussion of the identification of forms and numbers may be found in the *Metaphysics*, especially in XIII, 6–9 and XIV, 3–5.

were called numbers, and said to be built up from elements, and some things are discerned by intellect, others by knowledge, others by opinion, and others by sense perception, while the forms of the things are these numbers. And since the soul seemed to be able to produce motion and also able to produce knowledge in this fashion, some people have made it be intertwined out of the two together, declaring the soul to be a self-moving number.[4] 404b 30

There are differences about the sources of the soul—what they are and how many there are—especially among those who make them bodily, those who make them bodiless, and those who mix them and give an account of the sources as coming from both sorts. And there are differences too about the number of them, since some say there is one but others more than one; they give accounts of the soul in ways that follow from these opinions, for they assume, not unreasonably, that it is that among the primary beings which is by nature such as to cause motion. Hence, it seemed to some to be fire, since, among the elements, this is the one that has the smallest particles and is the most bodiless, and moreover is the one that primarily is moved and moves other things. Democritus has spoken in the most polished way in accounting for why each of these things is so, for he says that the soul and the intellect are 405a

4. This formulation of Xenocrates, a student of Plato, is discussed and criticized below, beginning at 408b 32. R. Catesby Taliaferro argues that the doctrine is older, deeper than Aristotle takes it to be, and reflected in some of Plato's dialogues. He takes the phrase as a metaphor equivalent to the myth of recollection, meaning that *human* souls are identical to one another in the way numbers are, and move themselves toward knowledge. See *The Timaeus and Critias of Plato*, Pantheon Books, Bolingen Series, 1952, p. 20.

405a 10 the same thing, while the latter is made of the first indivisible bodies, apt to be moved because of the smallness of the particles and their shape; of the shapes, he says that the spherical is the most easily moved, and that both the intellect and fire are of such a shape. Anaxagoras seems to speak of the soul and the intellect as different, as we said before, but he uses them both as a single nature, with the exception that he sets down the intellect especially as the source of all things; he claims at any rate that it alone among beings is simple, unmixed, and pure. But he accounts for both knowing and moving by the same source, saying that intellect set the whole in motion. Thales too, from the stories 405a 20 people recall of him, seems to have assumed the soul to be something that causes motion, if indeed he said the magnetic stone has a soul because it moves iron. But Diogenes, along with certain others, took the soul to be air, supposing this to be the thing with the smallest particles of all, and a source, on account of which the soul both knows and moves things, knowing insofar as it is the first thing out of which the rest are composed, and being able to produce motion insofar as it is the smallest. Heracleitus also said that the soul is the source of things, if indeed it is the vapor out of which he composes the other things, and is both the most bodiless thing and always flowing; and he said that what is in motion is known by what is in motion, both he and most people supposing that beings are in motion. 405a 30 And Alcmaeon seems to have assumed about the soul things just about the same as these, for he said that it is immortal because it is like the immortal things, and that this likeness belongs to it because it is always in motion, since he said that each of the divine things, the moon, 405b the sun, and the stars, as well as the whole heaven, are always continuously in motion. Among the cruder thinkers, some have even declared the soul to be water, as did Hippo; they seem to have been persuaded of this by the fact that the generative material of all things is liquid. For Hippo also refuted those who claimed the soul was blood by saying that the generative material is not blood, and that this is the first soul. Others, such as Critias, said it is blood, assuming perceiving to be the thing most properly belonging to soul, and this to be present because of the nature of blood. So each of the elements has gotten a judge in its favor, except earth; no one has declared in favor of it, unless someone has said the soul is composed of all the elements, or *is* all of them. 405b 10

Thus all thinkers, so to speak, define the soul by three things, motion, sense perception, and bodilessness, and each of these is traced back to the sources of things. Hence those who define it by its knowing make it either an element or something made of elements, saying just about the same thing as one another, with one exception. For they say like is known by like, and since the soul knows all things, they compose it out of all the sources. So as many of them as say there is some one cause and one element also set down the soul as one thing, such as fire or air, while those who say that the sources are more than one also make the soul more than one thing. Anaxagoras alone says that the intellect is not acted upon and has nothing in common with anything else. But how and through what cause it is going to know, if it is such, he has not said, nor is it clear from anything he has said. But those thinkers who put contraries into the sources of things also compose the soul out of contraries, while those who set down one of the contraries, such as heat or cold or some other such 405b 20 thing, as the source, also set down the soul similarly as some one of them. And in doing this they follow names, some saying that the soul is heat because life is so named on account of it, others that it is cold because it is called soul on account of breathing and cooling down.[5] These, then, are the things that have been handed down to us about the soul, and the reasons they have been said in these ways.

405b 30

Chapter 3 One ought first to examine what concerns motion, for perhaps it is not only false that the thinghood of the soul is of the sort that is meant by those who say it is what moves itself, or is capable of moving other things, but it is an impossible thing for motion to belong to it. That it is not necessary for what moves other things to be in motion itself has been said before. There are two ways in which all things are moved—for it is either in virtue of something else or in virtue of themselves; by being moved in virtue of something else, we mean whatever is moved on account of being in something that is moved, as sailors are, since they are not moved in the same sense as the boat. The latter is moved in its own right, and they by being in something that is moved. (And this is clear in the case of parts of a whole, since walking is a motion properly belonging to feet, and hence this also belongs to human beings, but this does not apply to the sailors.) So since being in motion is meant in two ways, we are now examining concerning the soul whether it is moved and has a share in motion in its own right.

406a

406a 10

5. The Greek words for life and soul resemble, respectively, verbs meaning to boil and to cool.

And since motion is of four sorts—change of place, alteration, wasting away, and growing—it is with one, or more than one, or all of these that the soul would be moved. And if it is moved in a way that is not incidental, motion would belong to it by nature; and if this is so, then also it would have a place, since all the motions mentioned are in place. But if the thinghood of the soul is to move itself, then being in motion would belong to it not incidentally, as it does to what is white or three feet long; for these are also in motion, but incidentally, since that to which they belong, the body, is in motion. And for this reason, there is no place belonging to these attributes, but there would be one belonging to the soul, if indeed it shares in motion by its nature. Besides, if it is moved by nature, then also it could be moved by force, and if by force, then also by nature, and the same thing holds also for rest; for in that place to which something moves by nature, it also rests by nature, and similarly, in that place to which it is moved by force, it also rests by force. But of what sort the forced motions and states of rest of the soul would be, it is hard to give an account, even for those who want to make up stories.[6] Also, if the soul is going to move upward, it will be fire, but if downward, earth, since these motions belong to these bodies, and the same argument also concerns the ones in between. What's more, since the body obviously moves, it is reasonable that the soul moves it with those

406a 20

406a 30

6. If the various myths in Plato's dialogues come to mind, that is probably what Aristotle intends, since his verb here puns on Plato's name. In particular, the cave image at the beginning of Bk. VII of the *Republic* is full of the language of forced motion, and is easily misunderstood, since, while the analogy is political, its meaning refers only to education.

motions by which it too is moved, but if this is so, then it is also true to state the converse, that the motions by which the body is moved are those by which the soul is too. But the body is moved by change of place, so that the soul would also change in ways that correspond to the body, either as a whole, or by an exchange of places among its parts; and if this is possible, the soul would also be capable of leaving the body to come back in again, from which would follow the rising back to life of the dead among animals.

406b

By incidental motion the soul could be moved even by something else, for an animal might be pushed by force, but it is necessary that something whose thinghood includes moving itself not be moved by anything else *except* incidentally, in the same way that something good in its own right, or by virtue of itself, cannot be good by virtue of or for the sake of something else, except incidentally. But one would say that the soul is moved by things it perceives most of all, if it is moved. But surely even if the soul itself moves itself, then at any rate it would be moved, so that, if every motion is a stepping outside itself of the thing moved insofar as it is moved, the soul would step outside its own thinghood, if it moves itself not incidentally, but motion belongs to its thinghood in its own right.[7] And some people say that the soul moves the body in which it is in the same way it moves itself, for instance Democritus, who says just about the same thing as the comic poet Philippus;

406b 10

7. To avoid this consequence, one would have to say that the soul is that which, *by* its very nature, moves or alters itself in ways that do *not* belong to its nature. Finding the nature of the soul in self-motion may sound impressive, but it seems to be either self-contradictory or empty.

for he says that Daedalus made his wooden Aphrodite move by pouring in molten silver, and Democritus speaks in a similar way in saying that the indivisible spheres, while moving, since they are of such a nature as never to be at rest, drag the whole body with them and set it in motion. But we shall ask whether this same thing produces rest, but how it could produce it is difficult or even impossible to say. And generally, it seems that it is not in this way that the soul moves the animal, but rather by means of some sort of choice and thinking. 406b 20

In the same way too, *Timaeus*[8] argues from nature that the soul moves the body, for by being in motion itself it also sets the body in motion on account of its being intertwined with it. For when the soul had been composed out of the elements and partitioned out according to the numbers of harmonic ratios, so that it might have an innate perception of harmony and carry around the whole in harmonious revolutions, the craftsman bent its straight shape into a circle, and, having taken the one circle apart into two, uniting them again at two places, he divided one of them into seven circles, as though the revolutions of the heaven were motions of the soul. Now first of all, it is not right to say that the soul is a magnitude, for by the soul of the whole, it is clear that he means something such as is what is called intellect. (For surely it is not any sort of perceiving or desiring power, since the motions of these are not around in a circle.) But the intellect is one and continuous in the same way as the act of thinking, and 406b 30 407a

8. A character invented by Plato. What follows is from the dialogue of the same name.

the act of thinking is identical with the things thought; these are one in the sense of one series, the way number is, but not in the way a magnitude is one. Therefore the intellect is not continuous in that way either, but 407a 10 is either without parts, or continuous in a way that no magnitude is.

For how would it even think, if it were a magnitude? Will it be by means of some one of its parts, and will the parts be portions of magnitude or like points, if one ought even to call these parts? Now if it thinks by something that corresponds to a point, and these are infinite, it is clear that it will never get through them, but if it thinks by a portion of magnitude, then it will think the same thing many times, or infinitely many. But it is obviously capable of thinking something once. But if it is sufficient for it to touch on something with any one of its parts, why does it have to move in a circle, or even have magnitude at all? But if it is necessary, in order to think anything, to touch it with the whole circle, what does the touching by its parts amount to? Again, how will it think what has parts by means of something without parts, or what has no parts by something with 407a 20 parts? It is necessary that this circle be the intellect, for the motion of the intellect is thinking and that of a circle is revolving; so if thinking is a revolving, then intellect would be the circle whose revolving is the act of thinking. But then what will it always be thinking (as it has to if the revolving is everlasting)? In thinking about matters of action there are limits (since all such thinking is for the sake of something else), while contemplative thinking is likewise bounded, by reasoned speech; and every reasoned articulation is either a definition or a demonstration. Demonstrations are from a starting point and also have some sort of end, the inference or conclusion (and if they do not conclude, still they do not at any rate turn back again to the beginning, but by always taking up another middle term and ending term they proceed in a straight line, but a revolving motion turns back again to the beginning), while all definitions are finite. Also, if the same revolution takes place many times, it would be necessary to think the same thing many times. 407a 30

Anyway, thinking seems like some sort of stillness or coming to rest rather than motion, and inference seems the same way. And surely what is not easy but forced is not supremely happy either; but if motion does not belong to the thinghood of the soul, it would be moved contrary to nature.[9] Also, for the soul to be mixed with the body and incapable of getting free would be burdensome, and in fact something to be avoided, if indeed it is better for the intellect not to be with the body, a thing that is both customarily said and subscribed to by many. And it is even unclear what the cause is for the heaven to be carried around in a circle; for the thinghood of the soul is not responsible for being carried in a circle, but it is moved in that way incidentally, nor is the body responsible, but the soul more so than it. But it is not said that it is better, even though this is why it was necessary for the god to make 407b 407b 10

9. At 34b of the *Timaeus*, the ensouled world is called "a blessed god." There is something wrong with the text of the second clause in the manuscripts, but Ross's decision to leave out the "not" makes it worse. *Timaeus* makes the thinghood of the soul thinking rather than moving, and if thinking seeks rest, adding an everlasting revolving to the soul would everlastingly go against its grain.

the soul be moved in a circle, because it is better for it to be moved than to stand still, and to be moved in this way rather than in another.[10]

But since such an examination is more suited to other writings, let us set this aside for now. But the following absurdity goes with both this account and most of those that concern the soul. They attach the soul to the body and set it into it, determining no further what the cause of this is or what the condition of the body is, and yet this would seem to be necessary, for by the partnership of soul and body the one acts and the other is acted upon, and the one is moved while the other moves it, but none of these things belongs to just any two things in relation to each other. But people put their effort into saying what sort of thing the soul is, while they determine nothing further about the body that receives it, just as though, in the manner of the Pythagorean myths, any random soul were to be clothed in any random body. For while each body seems to have its own proper look and form, they talk as if one were to say that carpentry is transmigrated into flutes; but the art has to use tools and the soul has to use the body.

407b 20

Chapter 4 There is a certain other opinion that has been passed down about the soul, not a bit less persuasive to many people than those that have been mentioned, though it has had to answer charges, as though

10. This is not Aristotle's own opinion, but Timaeus's fundamental principle (*Timaeus*, 29E); in the telling of his likely story, though, he fails to assign it to the motion of the soul. Aristotle's own account, in Bk. XII, Ch. 7 of the *Metaphysics*, has many similarities to Timaeus's, but decisive differences: the organized body of the world is moved by an intellect which is not a soul, is not in motion, and acts on the world only by final causality.

before inspectors, even in the writings that have become current among the public. People say that the soul is some sort of harmony, for they assert that harmony is a blending and combining of contraries, and that the body is composed of contraries. And yet harmony is some ratio or putting together of things that have been mixed or joined, and the soul cannot be either of these. Also, causing motion does not belong to a harmony, but everybody, so to speak, assigns this most of all to soul. And it fits better to speak of harmony in relation to health, and generally as belonging to bodily excellences, rather than in relation to the soul; this is clearest if one tries to explain the experiences and acts of the soul by means of harmony, for it is difficult to make it fit. Further, if we speak of harmony while keeping an eye on its two senses—its most governing sense being the joining of magnitudes that have motion and position, whenever they are so fitted together that they admit nothing of the same kind between them, and its next sense being the ratio in which things are mixed—in neither sense is it reasonable.[11] The claim that the soul is a joining of the parts of the body is especially easy to test. For the joinings of the parts are many and of many sorts; so of what or in what way ought one to

407b 30

408a

408a 10

11. The Greek word *harmonia* originally means the sort of joining, such as mortising, that is done in carpentry. The tendencies of the parts to pull apart are harnessed to hold the whole artifact together. The word extends to consonance in music by a close analogy. Tones that sound against each other can be tuned to produce the larger unity of a consonant interval. The Pythagoreans discovered that the string lengths that produce consonance, when other things such as thickness and tension are equal, are in the ratios of small whole numbers. It is appealing to stretch the metaphor again to apply to the unity of a living body, but Aristotle finds nothing left of the original analogy.

assume the intellect to be a joining, or the perceptive or desiring power? And it is similarly absurd for the soul to be the ratio of the mixture, for the mixture of elements according to which something is flesh does not have the same ratio as that according to which it is bone, so it would follow that a body has many souls in every part of it, if the parts are all made of mixtures of elements and the ratio of the mixture is a harmony and a soul.

408a 20 One might also demand an explanation of this from Empedocles, since he claims that each of the parts is a certain ratio. So is the ratio the soul, or is it rather that a ratio comes to be present in the parts because the soul is something else? Also, is friendship responsible for whatever mixture there happens to be, or for one that is according to a ratio, and if the latter, is friendship the ratio or something else besides the ratio? These opinions, then, have impasses of this sort. And if the soul is something other than the mixture, why in the world is it taken away at the same time as is whatever makes the material *be* flesh or be the other parts of the animal?[12] On top of these things, assuming there is not a soul for each of the parts, if the soul is not the ratio of the mixture, what is it that is destroyed when the soul departs?

408a 30 It is clear, then, from what has been said, that it is not possible for the soul to be a harmony, nor for it to be carried around in a circle. But it is possible for it to be moved incidentally, as we said, and even move itself,

12. Aristotle's wording is simpler: why is the soul taken away at the same time as is the being-flesh? At 429b 10–22, he makes clear that being-flesh is the intelligible aspect of flesh, regarded independently of its material.

in the sense that the body in which it is present moves, while this body is moved by the soul, but in no other way is it possible for it to be moved with respect to place. But one might more reasonably raise the question about it as being moved by looking to things such as these: 408b we say that the soul grieves or rejoices, is confident or afraid, and also is angry, as well as that it perceives and thinks things through, and all these things seem to be motions. Hence one might suppose that the soul is moved, but this is not necessarily so. For even if, as much as possible, grieving and rejoicing and thinking things through are motions, and with each of these something is moved, still the being moved is *by* the soul; for example, being angry or afraid is the heart's being moved in a certain way, and thinking things through is perhaps something of this sort or else of some other part, while certain of these make some parts move in 408b 10 place and others make other parts change by altering in quality (which ones and how is another story), yet to say that the soul gets angry is as if someone were to say the soul weaves cloth or builds a house. For it is better, perhaps, not to say that the soul pities or learns or thinks things through, but that the human being does these things by means of the soul, and this not in the sense that the motion is in the soul but that it sometimes goes up to the soul and sometimes comes from it; for example, sense perception comes from these things here, but calling something back to memory goes from the soul to the motions or stopping places in the sense organs.

But the contemplative intellect seems to be within while being an independent thing, and not to be destroyed. For it would be destroyed most of all by the 408b 20 dimming effect of old age, but it turns out in fact just as in the case of the sense organs; for if an old person were to get a certain kind of eye, he would see just as a young person does. Thus old age results from something's happening not to the soul, but to that in which the soul is, just as in bouts of drunkenness or in diseases. So too, contemplative thinking or intellectual insight wastes away because something else in us is destroyed, but is itself unaffected. But thinking things through and loving or hating are attributes not of the intellect but of that which has intellect, insofar as it has it. For this reason, when the latter is destroyed, the intellect neither remembers nor loves, for these acts did not belong to it but to the composite being which has perished; the intellect is probably something more divine and is unaffected.

408b 30 So it is clear from these things that the soul is not the sort of thing that can be moved. And if it cannot be moved at all, it is obvious that it does not move itself. But of the opinions mentioned, the most irrational by far is the one that says the soul is a self-moving number,[13] for impossibilities arise for those who say this, first of all as consequences of its being moved, but for them in particular from calling 409a it a number. For how ought one to think of numerical units as being moved, and by what, and by what means when they are without parts and undifferentiated? For insofar as they are such as to cause motion and such as to be moved, they have to have differences. Also, since they say that a line, when moved,

13. See 404b 27–30 and note.

produces a surface, and a point, when moved, produces a line, then also the motions of the numerical units would be lines, since a point is a unit having position, while the number that is the soul is already somewhere and has a position. Also, if one takes away a number or a unit from a number, a different number remains, but plants and many animals live on when divided up, and seem to have souls that remain the same in species. 409a 10

It would seem to make no difference whether one speaks of numerical units or of little bodies, for if the atomic spheres of Democritus were to turn into points, so long as some amount of them were to remain, there would still be among them some part that causes motion and another part that is moved, just as in a continuous magnitude. For whether they are said to cause motion or to be moved does not depend on a difference in magnitude or smallness, but is so just because there is an amount of them; hence it is necessary for there to be something that moves the units.[14] But if, in an animal, the part that moves it is the soul, this will also be the case in the number, so that the soul would not be the part that causes motion and the

14. The argument as applied to continuous magnitudes is in the *Physics*, 257a 33–258a 27. It would apply just the same to anything that is divisible in quantity, such as a collection of points or units. What it rests on goes to the heart of Aristotle's understanding of motion. A moving thing can bump something else along but can never *originate* motion; the origin of any motion must be something fully at-work in an unchanging way, and it would be self-contradictory for it to be simultaneously engaged in that same motion. Hence any apparently self-moving thing has parts, and whatever it is within it that causes it to move must be, with respect to that particular kind of motion, motionless.

part that is moved, but the part that causes motion only. But then how can this be a unit? For then there 409a 20 would have to be in it some difference in relation to the other units, but what difference could there be in a point that is a unit except position? But if the units in the body are something distinct from points, the units will still coincide with the points, for each of them will hold down the place of a point; however, if two things can coincide, what would prevent an infinite number? For those things of which the place is indivisible are themselves indivisible from one another. But if the points in the body *are* the number of the soul, or the number of points in the body is the soul, why wouldn't all bodies have souls? For there seem to be points in all of them, in fact an infinity of them. Also, how would it be possible for the points to be separated and set free 409a 30 from the bodies they are in, since at any rate lines cannot be taken apart into points?

Chapter 5 So this opinion turns out, as we said, to be saying the same thing as do those who set down the soul as some sort of finely divided body, though in another 409b way the manner in which Democritus says the body is moved by the soul is an absurdity peculiar to itself. For if the soul is present in the whole of the perceiving body, then necessarily there are two bodies in the same place, if the soul is some sort of body. But for those who call the soul a number, there are many points in one point, and every body has a soul, unless some different sort of number comes in that is something other than the points that are present in the body; and the animal turns out to be moved by the number in exactly the same way that we said Democritus moves it. For what difference does it make to speak of little spheres or big units, or of units being carried around at all? For in both versions it is necessary to move the animal by the being in motion of those things. So for those who intertwine motion and number into the same thing, these and many other such consequences follow. For it is impossible not only for the definition of soul to be of this sort, but even for such a thing to be an attribute of soul; this is clear if one tries to use this formulation to account for the things that the soul experiences and does, such as reasonings, perceptions, pleasures, pains, and everything else of that sort—as we said before, it is not easy even to get a clue out of these words. 409b 10

Three ways by which the soul is defined have come down to us; some people have declared it the thing most suited to cause motion by setting itself in motion, others the body that is most finely divided or most bodiless in comparison to the rest. These opinions have certain impasses and internal contradictions which we have gone just about all the way through, but it remains to examine the way in which the soul is said to be composed of elements. People say this in order that it may perceive and recognize each of the things there are, but many impossible things necessarily follow from this statement. For they set down that like recognizes like, as though setting down that the soul is the things it recognizes. But these elements are not the only things there are, but there are many others, or rather perhaps an infinite number—the things made out of the elements. So let it be possible for the soul to recognize and perceive the things out of which each of these things is composed; still, by what means will it recognize and perceive the composite whole, such as what a god, a human being, 409b 20 409b 30

410a

or flesh, or bone is? And it is also similar with any other composite thing whatever, for none of these has the elements in it any which way, but each according to some ratio and means of combining, as Empedocles says of bone,

> The earth, full of gifts, in the wide chests of its melting pots
> Took as its portion two of the eight parts from the shining primal water
> And four from the fire god Hephaestus, and they became white bones.

So it does no good for the elements to be in the soul unless the ratios and combinations are also going to be in it, for the soul would recognize each thing that is like it, but not bone or a human being, if these are not in it too. That this is impossible, there is no need even to say, for who would raise the question whether a stone or a human being is in the soul? And it is similar with good and not-good, and the same way with all the rest.

410a 10

What's more, since being is meant in more than one way (for it means in one way a *this*, but in another way how much or of what sort something is, or any other of the ways of attributing being that have been distinguished),[15] will the soul be composed of all of these or not? But elements do not seem to belong to all sorts of beings in common. Well then, is the soul composed only out of as many elements as are in independent things? But then how does it also recognize each of the

15. A list of eight so-called "categories" of being may be found at *Metaphysics* 1017a 22–27.

other sorts of being? Or will they say that there *are* elements within each kind of being, and sources peculiar to each, out of which the soul is composed? Then the soul would be an amount *and* a sort of attribute *and* an independent thing. But it is impossible that, out of the elements of an amount, it should come to be not an amount but an independent thing; but these as well as other such consequences follow for those who say the soul is composed of all the kinds of being. And it is also absurd to say that like is unaffected by like,[16] but that like perceives like and recognizes what is like by what is like, but they do set down that perceiving is a certain way of being affected and moved, and similarly with thinking and recognizing as well.

410a 20

What has just been said bears witness to the many impasses and inconveniences there are in saying, as does Empedocles, that each thing is recognized by means of bodily elements, and in addition, by the parts that are like it. For as many of the parts in the bodies of animals as are simply earth, such as bones, tendons, and hair, do not seem to perceive anything, and so do not perceive their like, and yet this belongs to what they say. Also, in each of the primal elements there would be more ignorance than understanding, for each of them would recognize one thing, but be ignorant of many, namely all the others. And it turns out, for Empedocles at least, that his god would be the most ignorant of all, since it alone would fail to recognize one of the elements—strife—while mortals would recognize them all, since each mortal is composed of them all. And

410a 30

410b

16. For example, stirring some water into a glass of water doesn't change it. This opinion is discussed at the beginning of Bk. I, Ch. 7 of *On Coming to Be and Passing Away*.

in general, why wouldn't all beings have soul, since each of them either is an element or is made out of one or more or all the elements? Each would necessarily recognize one or some or all of them. And one might also raise as an impasse what in the world it is that makes one thing out of the elements, for they seem to be like material, but the most governing thing is that which holds the elements together, whatever it might be; but it is impossible that anything should be more powerful than the soul and rule it, and more impossible still that anything should rule the intellect, for it is reasonable, in accordance with nature, that *this* comes first and is what governs, but they say that the elements are the primary beings.

410b 10

All of them, both those who say the soul is made out of the elements on account of its recognizing and perceiving beings, and those who say that it is the thing most suited to cause motion, fail to speak about every soul. For not everything that perceives is capable of motion (for there seem to be, among animals, certain ones that are stationary with respect to place, and yet this is apparently the only kind of motion the soul gives the animal.)[17] It is similar for those who produce the intellect and the perceiving power out of the elements, for plants appear to live without sharing in perception, and many of the animals have no power of thinking things through. But even if one were to concede these points, and grant that the intellect is a part of the soul, and similarly with the perceiving power, not even

410b 20

17. In the *History of Animals*, Aristotle mentions some kinds of oysters and sponges as examples. They are, of course, in motion in ways other than change of place, but that is the only kind of motion considered by these earlier thinkers.

in that way would he speak universally about every soul, nor about the whole of any one of them. This happens too with the account given in the so-called Orphic verses, for they say that the soul comes in from the whole when things breathe, carried by the winds, but it is impossible for this to happen to plants, or even to some of the animals, if not all of them breathe; but this has escaped the notice of those who have had this conception. (If one must make the soul out of the elements, there is no need to make it out of all of them, for one part of a pair of contraries is sufficient to discern both itself and its opposite; for by means of the straight we recognize both it and the crooked, since the ruler is the judge of both, though the crooked is not the judge either of itself or of the straight.)

410b 30

411a

Some people also say that the soul is mixed into the whole of things, which is perhaps why Thales supposed that all things are full of gods. But this has in it certain impasses; for why is it that the soul does not produce an animal when it is in the air or in fire, but does so in the mixture, even though they think it is better in these separate elements? (But one might also inquire why the soul in the air is better and more immortal than the soul in animals.) But something absurd or irrational follows on either choice, since to say that a fire or the air is an animal is more irrational, and not to say they are animals when they have souls in them is absurd. They seem to assume that the soul is in these elements because the whole is of the same kind as its parts, so it is necessary for them also to say that the soul is of the same kind as these parts, if it is by the taking away of something from the surrounding environment into the animals that the animals become ensouled. But if the air when divided up is of the same kind, while the

411a 10

411a 20

soul is not all of the same kind, it is clear that, if some one part of the soul were to become present out of the environment, another part would not become present. Necessarily, then, the soul is either all of the same kind or is not included in any and every part of the whole.

It is clear, then, from what has been said, that it is not on account of being composed of the elements that it belongs to the soul to recognize things, and also that the soul is not rightly or truly said to move itself. But since recognizing things does belong to the soul, as well as perceiving and having opinions, and also yearning and wishing and desires in general, while also motion with respect to place comes about in animals by means of the soul, as do also growth, a peak, and decay, is each of these present in the whole soul, and do we think and perceive and move, and do and experience each of the other things by means of all of it, or different things by different parts? And is living something that is in one of these parts, or in many, or in all, or is there some other thing responsible for it? Now some people say that it has parts, and thinks by means of one part but desires by means of another. But then what in the world holds the soul together, if it is by nature divided up? For it is surely not the body, for it seems rather to be the soul, on the contrary, that holds the body together; at any rate, when the soul has gone out of it, the body gives off vapors and rots away. So if some other thing makes it be one, that would be the thing most properly meant by soul. But about that thing in turn, it would be necessary to inquire whether it is one or has many parts. For if it is one, why is not the soul also one right from the start? But if it is divided, the argument will again ask what holds that together, and so will proceed to infinity.

411a 30

411b

411b 10

One might also raise as an impasse about the parts of

the soul, what power each is to have in the body. For if the whole soul holds together all of the body, it would be appropriate for each of its parts to hold together some part of the body. But this seems impossible, for what sort of part the intellect would hold together, and in what way, is difficult even to invent. And it appears that plants can live when divided, as can also, among animals, some of the insects, as though they had souls that are uniform in kind, even if not in number; for each of the two parts has perception and moves according to place for some time. And if they do not continue to do so, there is nothing strange about that, since each part does not have the organs to preserve its nature. But nonetheless, in each of the two parts, all the parts of the soul are included, and the divisions of the soul are alike, one to the other and each to the whole, like one another as though they were not separate parts originally, and like the whole soul as though it was not divided originally. On the other hand, it would seem that the source of life in plants is some sort of soul, for this alone is shared by animals and plants, and this is separated from the source of perception, while nothing has perception without this.

411b 20

411b 30

Introductory Note to Book II

The definition of soul in Book II, Chapter 1, is built out of a series of oppositions which contain, in compact form, the heart of all of Aristotle's thinking:

σῶμα *sōma*	**ψυχή** *psychē*
ὕλη *hulē*	**εἶδος, μορφή** *eidos, morphē*
δύναμις *dunamis*	**ἐντελέχεια** *entelecheia*
ὑποκείμενον *hupokeimenon*	**τὸ τί ἦν εἶναι** *to ti ēn einai* **οὐσία ἡ κατὰ τὸν λόγον** *ousia hē kata ton logon*

The body (*sōma*) is material (*hulē*) for the soul (*psuchē*) which is its invisible look (*eidos*) and shapeliness (*morphē*), because the body has being as a potency (*dunamis*) for the being-at-work-staying-itself (*entelecheia*) that forms it. Thus the body as body is an underlying thing (*hupokeimenon*), because what it keeps on being in order to be at all (*to ti ēn einai*), and its thinghood as it is revealed in speech (*ousia hē kata ton logon*), is its soul. The central notion, the soul of the soul, is *entelecheia,* a word Aristotle invented by combining *enteles* (complete) with *echein* (to be by actively staying in some condition, to hold on as such-and-such a thing) in such a way that the combination is a pun on *endelecheia* (continuity, persistence).

The soul is the being-at-work-staying-itself of the living thing in more than one sense: *first,* in constantly maintaining the body as the kind of thing it is, the soul is *nutritive* (*threptikē*), transforming material from the surrounding world into what is needed in order that it might be at all, and this is the common definition of all soul. But in any living thing, what is maintained and preserved is its kind, not only in the changes of its own body through time, but in offspring like itself, so that Aristotle views even this first potency of the soul as primarily *reproductive* (*gennētikē*). And what is maintained and reproduced is not just a certain kind of body but a certain kind of life; in animals, the nutritive function is subserved by a *perceptive* (*aisthētikē*) potency, which brings along with it an awareness of pleasure and pain that transforms the need to take in material into appetite and desire. These means to mere living determine the character of the life so lived, and transform its end from mere living to a life that puts to work all its powers. They also alter the nutritive relation to the animal's surroundings, the taking in of material without its form, since perception is an openness that takes in form without material, a being-at-work (*energein*) that is a being-acted-upon (*paschein*). Hence, in a certain way, the whole of the perceptible world is present in, at-work in, the animal soul, and part of its life rather than mere material that might underlie it. Having defined the soul, Book II begins to explore the variety of powers of the soul as:

θρεπτική (*threptikē*) nutritive

γεννητική (*gennētikē*) reproductive

αἰσθητική (*aisthētikē*) perceptive

ἐνεργεῖν (*energein*) being at-work

πάσχειν (*paschein*) being acted-upon

Book II

Chapter 1 Let this be our discussion of the things handed on about the soul by those who came before us. But let us go back again and, as though from the beginning, try to distinguish what the soul is and what articulation of it would be most common to all its instances. One of the most general ways of being we call thinghood; of this, one sort has being as material, which in its own right is not a *this*, but another sort is the form or look of a thing, directly as a result of which something is called a *this*, and the third sort is what is made out of these. Now the material is a potency, but the form is a being-at-work-staying-itself, and this in two senses, one in the manner of knowledge, the other in the manner of the act of contemplating.

412a

412a 10

The things that seem most of all to be independent things are bodies, and of these, the natural ones, for these are the sources of the others. And of natural bodies, some have life while others do not. By life we mean self-nourishing as well as growth and wasting away. So every natural body having a share in life would be an independent thing having thinghood as a composite [of material and form]. And since this is a body, and one of a certain sort, namely having life, the soul could not be a body, since it is not the body that is in an underlying thing, but rather the body has being as an underlying thing and material [for something else]. Therefore it is necessary that the soul has its thinghood as the form of a natural body having life as a potency. But this sort of thinghood is a being-at-work-staying-itself; therefore the soul is the being-at-work-staying-itself of such a body. But this is meant in two ways, the one in the sense that knowledge

412a 20

is a being-at-work-staying-itself, the other in the sense that the act of contemplating is. It is clear, then, that the soul is a being-at-work-staying-itself in the way that knowledge is, for both sleep and waking are in what belongs to the soul, and waking is analogous to the act of contemplating but sleep to holding the capacity for contemplating while not putting it to work. But in the same person it is knowledge that is first in coming into being; for this reason the soul is a being-at-work-staying-itself of the first kind of a natural body having life as a potency. But such a body is organized [i.e., has

412b

parts subordinated to the whole as instruments of it]. (And even the parts of plants are organs, though utterly simple ones, as the leaf is a covering for the peel and the peel for the fruit, while the roots are analogous to the mouth, since both take in food.) So if one needs to say what is common to every soul, it would be that it is a being-at-work-staying-itself of the first kind of a natural, organized body. And for this reason it is not necessary to seek out whether the soul and body are one, any more than with wax and the shape molded in it, or generally with the material of each thing and that of which it is the material; for even though *one* and *are* are meant in more than one way, the governing sense of each of them is being-at-work-staying-itself.[1]

412b 10

So what soul is has been said in general, for it is thinghood as it is unfolded in speech, and this is what such-and-such a body keeps on being in order to be at all. It would be as though some tool, such as an axe, were a natural body, since its being-an-axe would be the

---

1. Another way to translate *entelecheia* might be "holding together actively as a whole."

thinghood of it, and that would be its soul; for if this were separated from it, it would no longer be an axe, other than ambiguously. But in fact it is an axe, since it is not of this sort of body that the soul is the meaning and the what-it-is-for-it-to-be, but of a certain sort of natural body, one that has a source of motion and rest in itself. But one ought to consider what has been said as applying even to the parts of such a body. For if the eye were an animal, the soul of it would be its sight, since this is the thinghood of an eye as it is disclosed in speech (and the eye is the material of sight); if its sight were left out it would no longer be an eye, except ambiguously, in the same way as a stone eye or a painted one. 412b 20

Now one should take what applies to the part up to the whole living body, for there is an analogy: as part [of perceiving] is to part [of the body], so is perceiving as a whole to the whole perceptive body as such. And it is not the body that has lost its soul that is in potency to be alive, but the one that has it, and the seed and fruit are in potency to be certain sorts of bodies. So just as the act of cutting is for the axe and the act of seeing for the eye, so too is the waking condition a being-at-work-staying-itself, but as the power of sight is to the eye and the capacity of the tool to the axe, so is the soul a being-at-work-staying-itself, while the body is what has being in potency. But just as the eyeball and the power of sight are the eye, so here the soul and the body are the living thing. So it is not difficult to see why the soul, or at least certain parts of it, are not separate from the body, if the soul is of such a nature as to be divided, since for some parts of the body, the soul is the being-at-work-staying-itself of those parts themselves. Nevertheless, nothing prevents some parts from being separate, so long as they are not the being-at-work of 413a

any part of the body. But it would be difficult to see [why the soul is not separate from the body] if the soul were the being-at-work of the body in the way that the sailor is of the boat. But let this stand as marking off
413a 10 and sketching out in an outline what concerns the soul.

Chapter 2 Now since what is clear and more knowable by reason arises out of what is unclear but more obvious, it is necessary to try again to go on in this way about the soul. For the defining statement not only needs to make clear what something is, as most definitions do, but also needs to include and display the cause. As it is, the statements of definitions are like conclusions. For example, what is squaring? The equality of an equilateral rectangle to an oblong rectangle. But this sort of definition is a statement of the conclusion, while one who says that squaring is the finding of a
413a 20 mean proportional states the cause of the thing.[2]

---

2. In Bk. II, Prop. 14 of the *Elements of Geometry*, Euclid shows how to find a square equal to any given rectangle, but only in VI, 13 does he reveal why the method works. If a line is a mean proportional between the sides of a rectangle, the square on it must equal the rectangle. Mean proportionality is the notion through which the equality of the two figures becomes intelligible. It is thus a logical middle term between the subject "square" and the predicate "equals rectangle." Aristotle is fond of examples in which the logical middle term is also a mean or middle or intermediate thing in some other sense. In Bk. II, Ch. 2 of the *Posterior Analytics*, where he discusses this at more length, the thing to be defined is an eclipse, and the middle term is the celestial body that gets between two others and causes it. The squaring example occurs in a more complex form in Plato's *Theaetetus*, where Theaetetus finds the cause that stands behind the long list of things his teacher can only demonstrate one by one (147D–148B). In ancient mathematics it was considered to be a mark of a good demonstration that it not only prove that something is so, but show the cause through which it is so.

So we say, taking this as a starting point for the inquiry, that what is ensouled is distinguished from what is soulless by living. But living is meant in more than one way, and if any one alone of the following is present in something, we say it is alive: intellect, perception, moving and stopping with respect to place, and the motion that results from nourishment, that is, wasting away as well as growing. And for this reason all plants seem to be alive, since they evidently have in themselves this sort of power and source, through which they have growth and decay in opposite directions, for they do not just grow upward but not downward, but in both directions alike, and in every direction, all of them that are continually nourished and live for the sake of their ends, so long as they are able to get food. The latter capacity can be present in separation from the others, but the others cannot be present in separation from it in mortal beings. This is obvious in the case of plants, since no other potency of soul belongs to them. 413a 30

Life belongs to living things, then, through this source, but to an animal, first of all, through sense perception; for even those that do not move or change their places, if they have perception, we call animals and do not merely say they are alive. Now of perception, the kind that first belongs to them all is touch, and just as the nutritive power is able to be present separately from touch and all sense perception, so is touch able to be present separately from the other senses. (By the nutritive power, we mean the part of the soul of which the plants have a share; and it is obvious that animals all have the sense of touch.) Through what cause each of these things happens, we will say later. For now, let us say this much only, that the soul is the source of these things that have been mentioned and is defined 413b 413b 10 by them: by nutrition, by sense perception, by thinking things through, and by motion.

Whether each of these is a soul or part of a soul, and if a part, whether in such a way as to be separated only in speech or also in place, is for some of these not difficult to see, but some present an impasse. For just as in the case of plants, some parts obviously live when divided and separated from each other, as though the soul in them is one in each plant in the sense of being-at-work-staying-itself but is in potency more than one, 413b 20 so too we see it happen with other capacities of the soul in the case of insects that have been cut in half; for each of the two parts even has both perception and motion with respect to place, and if it has perception, also imagination and appetite, since where there is perception there is also pain and pleasure, and where these are there is necessarily also desire. But about the intellect, that is, the contemplative faculty, nothing is yet clear, but it seems that it is a distinct class of soul and that it alone admits of being separated from body, as the everlasting from the destructible. As for the remaining parts of the soul, it is clear from what has been said that they cannot be separate, as some people say, though it is obvious that they are distinct in speech. 413b 30 For being perceptive is different from being capable of having opinion, if perceiving is also different from having opinion, and similarly with each of the other parts mentioned. Further, some animals have all these capacities, others some of them, and still others only 414a one (and this makes the difference among animals); why this is so must be considered later. And it turns out much the same with the senses, since some animals have them all, others some of them, and still others the one that is most necessary, touch.

And seeing that the means by which we live and perceive is meant two ways, as is the means by which we know (by which is meant in one sense knowledge but in another the soul, since by means of each of these we say we know something), likewise also we are healthy in one sense by means of health, but in another by means of either some part of the body or the whole of it. Now of these, the knowledge or the health is a form, and a certain look, and an articulation in speech, and a kind of being-at-work of what is receptive, in the one case a being-at-work of what is capable of knowing, in the other a being-at-work of what is capable of being healthy (for it seems that the being-at-work of what is active is present in what is acted on and placed in a certain condition). So since the soul is that by which in the primary sense we live and perceive and think things through, it would be a certain sort of articulation and form, and not an underlying material. For thinghood is meant in three ways, as we said, of which one way is as form, one as material, and one as what is made of both, while of these the material is potency and the form is being-at-work-staying-itself, so since what is ensouled is made of both, it is not the body that is the being-at-work-staying-itself of the soul, but the soul is the being-at-work-staying-itself of some body.

414a 10

For this reason, those who think the soul neither has being without a body, nor is any sort of body, get hold of it well, for it is not a body but something that belongs to a body, and this is why it is present in a body and in a body of a certain kind, and those earlier thinkers did not think well who stuck it into a body without also distinguishing which bodies and of what sort, even though there is no evidence that any random thing admits just any random thing within it. And

414a 20

this happens in accord with reason, since the being-at-work-staying-itself of each thing naturally comes to be present in something that *is* it in potency and in the material appropriate to it. That, then, the soul is a certain being-at-work-staying-itself and articulation of that which has the potency to be in that way, is clear from these things.

Chapter 3 Now of the potencies of the soul, all of those that have been mentioned belong to some living things, as we said, while to others some of them belong, and to still others only one. The potencies we are speaking of are those for nutrition, perception, motion with respect to place, and thinking things through. And in plants the nutritive potency alone is present, while in other living things this is present along with the perceptive. But if the perceptive potency is present, then so is that of appetite, for appetite consists of desire and spiritedness and wishing, while all animals have at least one of the senses, that of touch, and in that in which sense perception is present there are also pleasure and pain, as well as pleasant and painful sensations, and where these are present so is desire, since this is an appetite for the pleasant. Besides, they have a perception of food, for touch is the sense that perceives food since all animals are nourished by what is dry or moist and warm or cold, of which the sense is touch, though incidentally it is perceptive of other things. For neither sound nor color nor smell contributes anything to nourishment, but the flavor that comes from food is one of the things perceived by touch. Hunger and thirst are desires for, in the former case, what is dry and warm, and in the latter, what is moist and cold, while the flavor is a way of making these pleasant. One must get clear about these

414a 30
414b
414b 10

things later, but for now let this much be said, that those living things that have touch also have appetite; it is unclear whether they must also have imagination, but this needs to be examined later.[3] And in some living things, in addition to these potencies, there is present also that for motion with respect to place, and in others also the potency for thinking things through as well as intellect, as in human beings and any other living things there might be that are of that sort or more honorable.

So it is clear that there could be a single account of soul in the same way as of geometrical figure; for neither in that case is there any figure aside from the triangle and those that follow in succession, nor in this is there any soul aside from the ones discussed. But even in the case of the figures, there could be a common account which would fit them all, but would be appropriate to none of them in particular, and similarly in the case of the souls that were discussed. Hence it would be ridiculous to inquire after the common account, both in the one case and in the other, which would not be the particular account of any thing there is, nor apply to any proper and indivisible kind, while neglecting an account that is of that sort. (For what applies to the soul is just about the same as what concerns geometrical figures, for always in the one next in succession there is present in potency the previous one, both in figures and in things with souls, as the triangle is in the 414b 20 414b 30

3. This is a modification of what was said above at 413b 22, that any animal, since it has perception, also has imagination. The doubt expressed here becomes an assertion below at 415a 10–11 that there are exceptions, and the ones Aristotle has in mind are mentioned at 428a 10–11. The difficulty is resolved in Bk. III, Ch. 11.

quadrilateral and the nutritive potency in the perceptive one.) Therefore, for each kind, one needs to inquire what the soul of each is, as for a plant, a human, and

415a

a wild animal. And why they are in this sort of succession must be considered. For without the nutritive potency there is no perceptive potency, while the nutritive is present in separation from the perceptive in plants. Again, without the sense of touch none of the other senses is present, but touch is present without the others, for many animals have neither sight nor hearing nor a sense of smell. And among animals with the perceptive potency, some have the potency for motion with respect to place while others do not. Last and most rare are reasoning and thinking things through; for in those destructible beings in which reasoning is present,

415a 10

all the other potencies are also present, while reasoning is not present in all animals, but some do not even have imagination, though others live by this alone. But about the contemplative intellect there is a different account. It is clear, then, that the account that deals with each of these potencies is also the most appropriate account of the soul.

Chapter 4 The one who is going to make an inquiry about these potencies must necessarily get hold of what each of them is, and then inquire on in this way about what has directly to do with them and the other things about them. But if one needs to say what each of them is, such as what the potency for thinking or perception or nutrition is, even before this one must say what thinking is and what perceiving is, for in an account, activities and actions come before the potencies for

415a 20

them. But if this is so, then even before that one needs to have examined the objects of them, needing first, for the same reason, to mark out what concerns, say, food, or the thing perceived, or the thing thought. So first one ought to speak about food and begetting offspring, since the nutritive soul belongs also to the other living things and is the first and most common potency of soul, by which life is present in them all. Its work is to beget offspring as well as to use food, since the most natural thing for a living thing to do, if it is full-grown and not defective, and does not have spontaneous generation,[4] is to make another like itself, for an animal to make an animal and a plant to make a plant, in order to have a share in what always is and is divine, in the way that it is able to. For all things yearn for that, and for the sake of it do everything that they do by nature. (That for the sake of which is twofold, referring to the one to which the activity belongs, but also to the one for which it is done.)[5] So since it is impossible for them to share continuously in what always is and is divine, since no destructible thing admits of remaining one and the same in number, each of them does share in it in whatever way it can have a share, one sort more and another less, enduring not as itself but as one like itself, that is one with it not in number but in kind.

415a 30

415b

---

4. This apparent exception to the natural pattern refers to various worms and shellfish, as well as to the mistletoe, that seem to arise without individual parents in mud, water, or decaying organic material. It is discussed in the *Generation of Animals*, 762a 37–763b 16. The discovery, by means of the microscope, of egg and sperm cells, eliminated this anomaly. This passage makes it clear that Aristotle would have welcomed such evidence.

5. For the most part, the final cause of everything a living thing does is the maintenance of itself in its active wholeness. If it has offspring, the end aimed at by its activity, and even the meaning of its "self," become twofold.

Now the soul is the cause and source of the living body. This is meant in many ways, but the soul is alike 415b 10 a cause in three distinct ways, for as that from which the motion is, that for the sake of which it is, and as the thinghood of ensouled bodies, the soul is the cause. That it is the cause in the sense of the thinghood is clear, for the thinghood is responsible for the being of everything, while the being of living things is life, and of this the cause and source is the soul. Also, it is the being-at-work-staying-itself that is the articulation of what has being in potency. And it is clear that the soul is the cause in the sense of that for the sake of which, for just as intelligence acts for the sake of something, nature too acts in the same way, and that for the sake of which it acts is its end. But the soul is such an end by nature in living things, since all natural bodies are instruments of the soul, the bodies of plants in just the same way as 415b 20 those of animals, as though having being for the sake of the soul; and the soul is that for the sake of which they are in the twofold sense of being that to which they belong and that for which their actions are. But surely also the soul is the first thing from which their motion with respect to place comes, though this potency does not belong to all living things; but alteration and growth also come from the soul. For sense perception seems to be a kind of alteration, but nothing that has no share of soul has sense perception; and it is the same with growth and wasting away, since nothing that does not 415b 30 nourish itself either wastes away or grows naturally, and nothing nourishes itself which does not share in life.

And Empedocles has not spoken well in adding to this that growth happens to plants when they take root downward because earth moves that way by nature, and when they spread upward because fire moves that way. He doesn't understand up and down well either (for up and down are not the same for all things as they are for the whole cosmos, but as the head is for animals, so are the roots for plants, if one ought to speak of organs as different or the same by the work they do) but beyond this, what is it that holds the fire and earth together when they move in opposite directions? For they will be torn apart if there is not something that prevents it, and if there is, this is the soul, and it is responsible for the growing and feeding. But some people think that the nature of fire is by itself the cause of nutrition and growth, for it is evident that it alone of bodies feeds itself and grows, for which reason one might suppose that in plants and animals too, this is the thing that is at work. But though it is in some way jointly a cause it is surely not simply the cause, but rather the soul is, for the growth of fire goes on without limit, so long as there is something burnable, but all things put together by nature have a limit and proportion of size and growth, and this belongs to the soul, not to fire, and to the articulation of the meaning more than to the material.

416a

416a 10

But since the same potency of the soul leads to both nutrition and begetting it is necessary first to make distinctions about nourishment, since this potency is set apart from the others by this work. It seems that something contrary is food for its contrary, but not every contrary thing for every other, but all those contraries that have not only their coming-into-being but also their growth from each other; for many contraries come into being from each other, but they are not all of a certain

416a 20

amount—for instance something healthy comes from something sick. And it is evident that not even those that are of a certain amount are nourishment for each other in the same way, for liquid [such as oil] feeds a fire, but fire is not food for a liquid. Now among the simple bodies it seems most of all to be the case that one contrary is food and the other is fed, but there is an impasse, since some people say that like is nourished by like, in the same way that something grows, while to others it seems just the opposite, as we were just saying, that contrary is nourished by contrary, since like is unaffected by like, while food needs to be changed and digested, and for everything change is into an opposite or something in between. Also, the food is acted upon in some way by the thing fed, not the latter by the food, just as the carpenter is not acted upon by his material but it by him; the carpenter changes only from idleness into being at work. But it makes a difference whether the food is the final thing added to something or the first. If both are food, but the latter is undigested while the former is digested, then food would admit of being spoken of in both the ways discussed; for insofar as it is undigested, contrary is fed by contrary, but insofar as it is digested, like is fed by like. Therefore it is clear that both parties in a certain way speak both rightly and not rightly.

416a 30

416b

But since nothing is fed which has no share in life, it would be the ensouled body that is fed, insofar as it is ensouled, so that the food too is related to what is ensouled, and not incidentally. And being food is different from being something that produces growth; for insofar as the ensouled thing is of a certain amount, what is added produces growth, but insofar as the

416b 10

ensouled thing is a *this* and an independent thing, what is added is food (for it preserves the independent thing, which has being just for so long as it is fed), and it is productive of coming-into-being, not of the thing fed but of one like the thing fed, since the thinghood of the thing fed is present already, and nothing itself generates itself, but it does preserve itself. Therefore, this sort of potency of the soul is a source such as to preserve the thing's holding-on as the sort of thing it is, and the food gets it ready to be at work; for that reason, when deprived of food it cannot *be*. And since there are three things—the thing fed, that by which it is fed, and the thing that feeds it—the thing that feeds it is the first sort of soul, the thing fed is what has this soul, and that by which it is fed is the food. And since the right way to name all things is by their ends, while the end is begetting one like itself, the first sort of soul would be the potency for something to beget one like itself. 416b 20

Now that by which something is fed has two senses, just as that by which the helmsman steers refers to both his hand and the rudder, the former both causing motion and moved, the latter moved only. But all food needs to be able to be digested, and digestion is accomplished by something hot; for this reason every ensouled thing has heat. So what food is has been said in outline, but it must be clarified later in discussions devoted particularly to it.[6] 416b 30

Chapter 5 Now that these things have been distinguished, let us speak about all sense perception in

6. More on this topic can be found in *On Sleep and Waking*, Ch. 3.

common. And perception follows from being moved and acted upon, as has been said, for it seems to be a kind of alteration. And some say also that like is acted

417a

upon by like, but how this is possible or impossible has been spoken of in the accounts of acting and being acted upon in general.[7] And there is an impasse about why no perception takes place of the perceiving sense organs themselves, and why they do not produce a perception without things outside them, though they have in them fire and earth and the other elements of which there is perception, either of them in their own right or of attributes of them. But then it is clear that the perceptive power does not have being as a being-at-work but only as a potency, and this is why the sense organ is not perceived, just as what is burnable is not burned itself by itself without something to set it on fire, for then it would set itself on fire and there would be no

417a 10

need of a fire that was at-work-staying-itself. But since we speak of perceiving in two senses (for what has the potency of hearing and seeing we say hears and sees even if it happens to be asleep, as well as what is already at work hearing and seeing), even the power of perception should be spoken of in two senses, the one as being in potency, the other as being at-work, and similarly the thing perceived means both what is in potency to be perceived and what is at-work being perceived. So

---

7. See *On Coming to Be and Passing Away*, Bk. I, Ch. 7. Things have to be alike in some general way for interaction to be possible, but contrary in some more particular way. Part of the water in a glass doesn't act on another part because there is no contrariety, but the color of the glass doesn't act on the water either, because it is too unlike it.

first, let us speak of being acted upon and being moved as though they were the same as being-at-work, for motion is a kind of being-at-work, though not in the full sense, just as is said in other places.[8] Still, everything is acted upon and moved by something capable of acting and already being at-work. Hence it is acted upon in one sense by what is like but in another sense by what is unlike, as we said, for what is unlike is acted upon, but in the state that results from being acted upon it is like. 417a 20

But one must divide up the senses of potency and being-at-work-staying-itself; so far we have been speaking about them as unambiguous. There is something that has knowledge in the way that we say any human being is a knower, because humanity is part of the class of what knows and has knowledge, but there is also a sense in which we mean by a knower the one who already has, say, grammatical skill; and each of these is in potency but not in the same way, but the former is because his kind and his material are of a certain sort, while the latter is because he is capable of contemplating when he wants to, if nothing outside him prevents it. But the one who is already contemplating is a knower in the governing sense, since he is at-work-staying-himself knowing, say, this letter A. The first two, then, are both knowers in potency, but the former of them becomes so in activity when he has been altered by learning and has changed often from the contrary condition, while the latter does by 417a 30

8. *Physics*, 201b 31–33: "Motion seems to be a certain sort of being-at-work, but incomplete, and the reason is that what is in potency, of which motion is the [complete] being-at-work, is itself incomplete."

417b changing, in a different way, from having grammatical or arithmetical skill but not being at work with it, into being at work.

But "being acted upon" is not unambiguous either; in one sense it is a partial destruction of a thing by its contrary, but in another it is instead the preservation, by something that is at-work-staying-itself, of something that is in potency and is like it in the way that a potency is like its corresponding state of being-at-work-staying-itself. For the one who has knowledge comes to be contemplating, and this is either not a process of being altered (since it is a passing over into being oneself, namely into being-at-work-staying-oneself), or is a different class of alteration. This is why it is not right to say that a thinking being, when it thinks, is altered, any more than a housebuilder is altered when

417b 10 he is building a house. So the leading of one who contemplates and thinks into being-fully-at-work from being in potency is not teaching, but it is right for it to have a different name given to it; and the one who, out of being in potency, learns and acquires knowledge by the action of one who is fully at work and is disposed in the way we call teaching, either ought not to be said to be acted upon, or one must say there are two ways of being altered, the one a change to a condition of lacking something, the other a change to an active condition and into a thing's nature.

In the potency for perception, the first change comes about by the action of the parent, and when the living thing is born it already has what it takes to perceive, just as it has the capacity for knowledge. Being-at-work perceiving is described in the same way as contemplating,

417b 20 but differs in that the things that produce the being-at-work of perceiving are external, the visible and audible things, and similarly with the rest of the senses. The reason is that active perception is of particulars, while knowledge is of universals, which are in some way in the soul itself. Hence thinking is up to oneself, whenever one wishes, but perceiving is not up to oneself, since it is necessary that the thing perceived be present. And similarly too, even the kinds of knowing that deal with perceptible things are not up to oneself, and for the same reason, that the perceptible things are among particular, external things.

But there will also be an opportunity to get clear about these things afterwards; for now let this much be distinguished, that being in potency is not meant in an unambiguous way, but in one way as we might say a child is capable of being a general, and in another as we might say the same of one who is in the prime of life, and so too is it with the potency for perception. But since these distinct senses are without names, though it has been marked out that they are different, and how they differ, it is still necessary to use such words as "be acted upon" and "be altered" as though they were appropriate. And the perceptive being is, in potency, such as the perceived thing already is in full activity, as was said. So it is acted upon when it is not like the perceived thing, but when it is in the state that results from being acted upon, it has become likened to it, and is such as that is.

417b 30

418a

Chapter 6 One must speak first, for each of the senses, of the perceptible things. But the perceptible thing is meant in three ways, in two of which we say that we perceive the thing in its own right, but in one that we perceive it incidentally. Of the two, one is proper to each sense, the other common to them all. By proper I mean

418a 10

what does not admit of being perceived by another sense, and about which it is not possible for the sense to be deceived, as sight with color and hearing with sound and taste with flavor—touch has more than one distinct thing proper to it—but each sense distinguishes these and is not deceived that something is a color or a sound, but only about what or where the colored or sounding thing is. Things of that sort, then, are said to be proper to each sense, but the common things are motion, rest, number, shape, and size, for things like that are proper to none of them but common to them all.

418a 20

For a particular motion is perceptible by both touch and sight. A thing is said to be incidentally perceptible, for example, if the white thing is the son of Diares, for this latter is perceived incidentally, because it is incidental to the white, that is perceived, for which reason nothing is acted upon by the incidentally perceived thing as such. But of the things perceived in their own right, the proper ones are perceptible things in the governing sense, and the thinghood of each sense is by nature directed toward them.

Chapter 7 That to which sight is directed is the visible, and this is both color and something which one can put in words, though it happens to be without a name; what we mean will be clear to those who read on.[9] What is visible is color, and this applies to

418a 30

that which is visible in its own right, not in virtue of the kind of thing it itself is in its own definition, but because it has in itself the cause of its being visible.

9. At 419a 2–5, Aristotle mentions various things we would call "phosphorescent."

And every color has the potency to set in motion what is actively transparent, and this is the nature of color; and for that reason it is not visible without light, but of each thing every color is seen in light. Hence one must first say what light is. Now there is such a thing as the transparent. By transparent I mean what is visible but not visible in its own right, to put it simply, but on account of the color of something else. Air and water and many solid bodies are of this sort; it is not insofar as it is water or air that each of them is transparent, but because there is some nature inherent in them that is the same in both of them and in the everlasting body in the upper region of the cosmos. And light is the being-at-work of the transparent as transparent; and in that in which it is, there is, in potency, darkness. Light is a sort of color of the transparent, whenever it is at-work-staying-transparent by the action of fire or something of that kind, such as the body in the upper region, for something present in this is one and the same with fire. 418b 418b 10

What, then, the transparent is, has been said, and what light is, that it is not fire nor any body at all, nor anything that flows out of any body (for that would be some sort of body and of the same kind), but the co-presence of fire or any such thing in the transparent. For two bodies could not be in the same thing at the same time, and also light seems to be contrary to darkness, and since darkness is the absence of this sort of active condition from the transparent, it is clear that the presence of it is light. And Empedocles is not right, nor is anyone else who might also have said that light travels and at some time comes to be in between the earth and what is around it, though this escapes our notice; for this runs counter to the light of reason as well as to the appearances. Over a small interval it might escape 418b 20 notice, but that it goes unnoticed from where the sun rises to where it sets is asking too much.[10]

It is the colorless that is receptive of color, and the soundless that is receptive of sound. But the colorless is the transparent, as well as what is either invisible or barely visible, as the dark seems to be; and the latter is 418b 30 the way that the transparent is, though not when it is at-work-staying-transparent, but when it is transparent in potency, since the same nature is at one time dark 419a and at another time light. Not all visible things are in light, though, but only the proper color of each thing; but there are some things that are not seen in the light but produce perception in the dark, such as those fiery and shining appearances (which do not have a single word as a name), for example fungus, [decaying] flesh, fish heads, fish scales, and fish eyes. But it is not the proper color of any of these that is seen. Why these are seen needs another account, but for now this much is clear, that what is seen in light is color (and hence 419a 10 it is not seen without light, since this was the very being-color of it, its having the potency to move the transparent into being-at-work) and that the being-at-work-staying-itself of the transparent is light.

Here is a clear sign of this: if one puts something having color up against the eye itself, it is not seen.

10. From the moment of first light in the morning we see as much illumination miles to the west as we do to the east. The air above the horizon seems to have changed its condition all at once. There would need to be some good reason to believe that the light sneaks by us too fast to be noticed, since that is so unlike anything else in our experience. It is not credible as a mere postulate, and was still denied in the seventeenth century by Descartes. The first solid evidence that light travels was discovered in that century by Ole Römer, by telescope observation of eclipses of the moons of Jupiter, and a brilliant interpretation of them.

Rather, the color sets a transparent thing, such as air, in motion, and by this, if it is continuous, the sense organ is moved. For Democritus does not speak of this rightly when he supposes that if what is in between were empty, and if an ant were in the sky, if would be clearly seen; but this is impossible. For seeing comes about when what is capable of perception is acted upon by something, and since it is impossible that it be acted upon by the color itself that is seen, what remains is for it to be acted upon by what is in between, so that it is necessary that something be in between; and if what is in between were to become empty, it is not that nothing would be seen clearly, but nothing would be seen at all. 419a 20

Why, then, color needs to be seen in light, has been said. But fire is seen both in the dark and in the light, and this is so necessarily, since it is by its action that the transparent becomes transparent. The same account also applies to sound and smell, for no sound or smell produces perception when it is touching the sense organ, but by the action of the smell or sound what is in between is set in motion, and by this medium the respective sense organ is moved; but when one has put the thing that is sounding or giving off a smell up against the sense organ itself, it will not produce any perception. It is similar with touch and taste, though this is not apparent; the reason why will be clear later. But the medium of sound is air, while that of smell is nameless, for there is some common attribute of air and water, and just as the transparent is for color, so is this attribute that is present in both of these for what has smell, for even the animals that live in water seem to have a sense of smell. But a human being, or any of the animals on land that breathe, is unable to smell unless 419a 30 419b it is breathing. The reason for this will also be discussed later.

Chapter 8 And now let us first make some distinctions about sound and hearing; sound is of two sorts, one a certain way of being-at-work, the other a potency, for we say of some things that they do not have a sound, such as a sponge or wool, but of others that they have one, such as bronze and whatever is solid and smooth, because they are capable of sounding (that is, of making there be sound at-work between themselves and what hears). The sound at-work comes about always from something, against something, and in something, for it is the striking of a blow that produces it. Hence it is impossible for a sound to come about from any single thing, for the thing striking and the thing struck are distinct, so that the sounding thing sounds against something, and the striking of a blow does not happen without motion from place to place. And as we said, sound is not the striking of any random things, for wool makes no sound by being struck, but bronze and all smooth and hollow things do; the bronze makes a sound because it is smooth, and the hollow things because they produce many blows after the first one by a bouncing back and forth, since what is set in motion is unable to get out. Further, sound is heard in air, and, though less so, in water, although it is not the air or water that is chiefly responsible for sound, but there has to be a striking of solid bodies against one another and against the air. This happens when the air that is struck remains in place and does not spread out; hence if it is struck quickly and forcefully, there is sound, for the motion of the thing striking it must outrun the yielding

419b 10

419b 20

of the air, as if one were striking at a heap or swirling ring of sand moving rapidly along.

An echo comes about when some air has become one mass on account of an enclosure that surrounds it and prevents its dispersal, and other air is bounced back again from it like a ball. It is likely that an echo always occurs, but not distinctly, since this would follow in the case of sound just as in that of light; for light is always reflected (for otherwise there would not be light everywhere, but darkness outside whatever is in the sunshine), though it does not reflect in the same way it does off of water or bronze or any other smooth surface, so as to produce a shadow by which we demarcate the light. It is rightly said that the void is the thing chiefly responsible for hearing, since what seems to be void is air, and *this* is what produces hearing, when it is set in motion as something that holds together and is one mass; but since air is apt to fly apart, it does not make itself heard if the thing that strikes it is not smooth. But in that case it becomes one mass all at once on account of the surface, since the surface of a smooth body is itself one.

419b 30

420a

What is capable of producing sound, then, is what is capable of moving air that is one continuously all the way to that which hears, but there is air naturally present as part of the organ of hearing, and since that is also in the air, when the outside air is set in motion, the inside air is moved. That is why the animal does not hear with every part of its body, nor does the air spread through it everywhere, for even the part that is going to be moved and filled with sound does not have air all through it. So the air itself is without sound, since it is easily dispersed, but whenever it is prevented from being dispersed, the motion of it is sound. But

420a 10 the air within the ears has been walled in so as to be stopped from moving, in order that there might be an accurate perception of every distinct sort of motion. For these reasons we hear even in water, because it does not get into the air itself that is naturally present, nor even inside the ear, on account of its spirals; when this does happen, the ear does not hear, nor does it when the eardrum is damaged, just as the eye does not see when the membrane on the eyeball is damaged. But a sign of whether the ear is hearing or not is that there is always a murmuring in it, as in a horn, for the air that is in the ears is always moved with a certain motion that belongs to it, but sound is from an external source and not from its own motion. This is also why people say that hearing is by means of something empty and resounding, because we hear by means of something that has air contained in it.

420a 20 Is it the thing that strikes or the thing that is struck that makes the sound? Or is it both, but in different ways? For sound is motion of that which can be moved in the same manner as things that bounce off smooth surfaces when one strikes them, so as was said, not everything that strikes and is struck makes a sound, as when a pin strikes a pin, but the thing struck has to be uniform, so that the air bounces and vibrates as one mass. The differences among sounding bodies are evident in the sound in its operation, for just as colors are not seen without light, neither is there a sharp or flat tone without sound. These are spoken of by way 420a 30 of metaphors from tangible things, for the sharp tone moves the sense organ a great deal in little time, but the flat tone moves it a little in much time. The sharp tone is not fast nor the flat tone slow, but the motion that produces the one or the other comes to be of that tone because of its swiftness or slowness, and this seems to have an analogy to the sharp and the blunt that concern touch, for the sharp is the sort of thing that stabs, while the blunt is the sort that presses, since the former causes motion in little time, the latter in much, so that the one is incidentally fast and the other is incidentally slow. 420b

So let what concerns sound have been marked out in this way. A voice is a sound belonging to something with a soul, for while nothing without a soul has a voice, such things as a flute or a lyre are said to have a voice by way of a likeness, as is any other thing without a soul that has a scale of sounds with pitches and articulation, and this is appropriate because the voice also has these. Many of the animals do not have voices, such as the bloodless ones, and, among those with blood, the fish (and this is reasonable if sound is a certain kind of motion of air), but fish that are said to have voices, such as those in the river Achelous, make a noise with their gills or some other such part, while a voice is a sound of an animal but not by any random part.[11] But since everything that makes a sound does so because something strikes something in something, and this last is air, it is reasonable that only those things would have voices that take in air. For nature uses the air that has already been inhaled for two jobs, just as it uses the tongue for both taste and speech, of which one, taste, 420b 10

11. The modern distinction of vertebrates and invertebrates corresponds roughly to Aristotle's primary distinction of animals as blooded or bloodless; the latter include insects, shellfish of all sorts, octopi and squid, and other species. (See Bk. IV, Ch. 5–9, of the *Parts of Animals*.) The Achelous, a large river in western Greece, had in it a kind of fish that made a noise by the rubbing of it gills; in the *History of Animals*, Bk. IV, Ch. 9, Aristotle distinguishes this sort of sound from the one the dolphin makes, which he considers a voice.

is necessary (and hence belongs to a greater number of
420b 20 animals), while recognition of meaning is for the sake of well-being; so too, nature uses breath as something necessary for internal warmth (the reason for which will be spoken of in other places)[12] and also for the voice so that well-being might be present.

The organ for breathing is the upper throat, and the part for the sake of which this is present is the lung, for it is by means of this part that land animals have more warmth than other animals. Breath is needed for the region about the heart primarily; hence the air must go inside by being breathed. And so the voice is the striking against the so-called windpipe of the air that has been breathed in, by the action of the soul in these parts (for
420b 30 not every sound of an animal is a voice, as was said—for it is also possible to make a noise with the tongue or in the way people do when they cough—but it is necessary for the part that causes the striking to have soul in it and some sort of imagination with it, since the voice is some sort of sound that is capable of carrying a meaning), and it is not, like a cough, a striking by
421a the air that is breathed in, but by means of this the animal makes the air in the windpipe strike against the windpipe. A sign of this is the inability to speak when one is inhaling or exhaling, rather than holding air in, since the one who holds air in causes motion with it. It is also clear why fish are without voice, since they have no windpipes; they do not have this part because they do not take in air or breathe. Why they do not is another story.

12. One such place is the beginning of III, 6, of the *Parts of Animals*, where the question at the end of this chapter is also addressed.

Chapter 9 About smell and what it perceives it is less easy to draw distinctions than about the senses that have been spoken of, for it is not as clear what sort of thing a smell is, as it is with a sound or a color. The reason is that we do not have precision in this sense, but are inferior to many animals; for human beings smell poorly, and do not even perceive any smells that are without pain or pleasure, as though the sense organ is without precision. It is a reasonable supposition that animals with hardened eyes perceive colors in that same way, and that there are no distinctions of colors for them except as frightening and not frightening; the human race is that way concerning smells. And while there seems to be an analogy with taste, and the kinds of flavors are similar to the kinds of smells, still we have a more precise sense of taste because it is a certain type of touch, and that is the most precise sense a human being has. For in the other senses, the human being is left behind by many of the animals, but with respect to touch he is precise in a way that greatly surpasses the rest, and this is why he is the most intelligent of the animals. A sign of this is that within the human race, being naturally well or badly endowed with intelligence depends on the organ of this sense and not on the others, for those with tough skin are badly equipped by nature for thinking, but those with tender skin are well equipped.

421a 10

421a 20

And just as one flavor is sweet, another bitter, so too are smells, but while some things have a correspondence in smell and flavor—I mean, for example, a sweet smell and a sweet flavor—others are opposite. And similarly, a smell is pungent, astringent, acidic, or oily. But since, as we were saying, smells are not as strongly distinguishable as flavors, they have gotten their names

421a 30

421b

from likenesses to the things that have them, a sweet one being a saffron or honey smell, a bitter one either a thyme smell or some such thing, and the same way with the rest. And just as with hearing and each of the other senses, there is something audible and something inaudible, or something visible and something invisible, so too with the sense of smell there is something odorous and something odorless. A thing is odorless either on account of being entirely incapable of having a smell, or on account of having one that is slight and insignificant. In the same ways, a thing is spoken of as tasteless.

And the sense of smell also acts through a medium, 421b 10 such as air or water; for the things that live in water also seem to have perception of smell, the ones with blood and the bloodless ones alike, just as do the things that live in air, since some of them even move toward their food from far away when they become aware of a scent. And hence there is obviously an impasse, if all things perceive smell in the same way, and if a human being does so when inhaling, but does not perceive smell when he is not inhaling but exhaling or holding his breath, not from far away nor from nearby, nor even in case something is placed inside his nostril. Being unable to perceive what is placed on the sense organ itself is something common to all animals, but not perceiving something without inhaling is peculiar 421b 20 to human beings; this is clear to those who try it out. So the bloodless animals, since they do not breathe, would have some other sense besides the ones that are spoken of; but this is impossible if it is smell they are perceiving, since the perception of a scent, either malodorous or fragrant, is the sense of smell. Also, they are obviously destroyed by those strong smells by which a human being is, such as those of tar, sulfur, and such things.[13] Necessarily, then, the bloodless animals perceive smell, but not by breathing.

It seems likely that this sense organ differs in human beings from that of the other animals, just as human eyes differ from those of animals with hardened eyes; the human ones have eyelids as a screen, like a casing, and do not see without moving or raising them, while the animals with hardened eyes have nothing of that sort, but immediately see what comes into the transparent medium. So too, then, the organ of smell in some animals is uncovered, as is the eye, while in those that take in air it has a covering that uncovers it when the animal inhales, by widening the passageways and openings. And this is why the animals that breathe do not perceive smells in water, for they must breathe in order to smell, and they cannot do this in water. Smell belongs to what is dry (as taste belongs to what is moist), and the organ of smell is dry in potency.[14]

421b 30

422a

Chapter 10 What is perceived by taste is a certain sort of tangible thing, and this is the reason why it is not perceived through a distinct body as a medium, since that perceived by touch is not either. And the body in which flavor, the thing perceived by taste, is

422a 10

13. Strictly, it is not the animal but only the sense organ that is destroyed by such excesses, except in the case of touch. See below, 435b 7–19. On what the bloodless animals are, see the note to 420b 10, above.

14. This cryptic remark is expanded in *On Sense Perception and Perceptible Things*, Ch. 5. The medium of smell is always moist, and so is the organ of smell (425a 5), but it is affected by the dry content alone; the way in which taste differs is taken up in the following chapter.

present, is in something moist as its material, and this is a tangible thing. Hence, even if we lived in water we would perceive that something sweet was thrown into it, but our perception of the sweet would not be through a medium, but by its being mixed with moisture, just as in a drink. Color, on the other hand, is not seen by being mixed in that way, nor by means of things flowing out of bodies; with taste, there is nothing in the role of a medium, but just as color is what is seen, flavor is what is tasted. But nothing produces a perception of flavor without moisture, but such a thing has moisture either actively or in potency, something salty, for example, since it is itself easily dissolved and adapted to the dissolving action of the tongue. 422a 20

And just as sight is of both the visible and the invisible (for darkness is invisible, but sight discerns this as well), and also of what is too bright (for this also is invisible, though in a different way from darkness), and hearing similarly is of both sound and silence, the one audible, the other not audible, as well as of sound of great magnitude, as with sight and what is bright (for just as a small sound is inaudible, in a certain way one that is great and violent is too), while "invisible" is meant in one sense completely, as in the case of other sorts of impossibilities, but in another sense of what has little or no capacity to be seen, though it is of such a nature as to be seen, as with a footless animal or unpitted fruit—so too taste is of both what has taste and what does not, the latter including what has a slight or bad flavor and what is destructive of the sense of taste. A first distinction seems to be the drinkable and the undrinkable (for some sort of taste belongs to both, but to the latter a taste that is bad and destructive, to the former one that is in accord with nature), and the 422a 30 drinkable belongs to touch and taste in common. And since what is perceived by taste is moist, its sense organ must not be either at-work staying moist or incapable of becoming moist, for the sense of taste is affected in a certain way by the thing that is tasted, in that respect in which it is something tasted. Therefore, the sense organ suited to tasting, since it must be moistened, has to be capable of surviving being moistened, but not be moist. A sign of this is that the tongue has no perception either when it is dried out or when it is too moist, which it becomes by contact with a first moisture, as when someone who has first tasted some strong flavor then tastes another one, or in the case in which all things seem bitter to sick people because they taste them with a tongue full of that sort of moisture.

422b

422b 10

The kinds of flavors, as also in the case of colors, are unmixed contraries, the sweet and the bitter, with the oily bordering on the former and the salty on the latter, and between these the pungent, the astringent, the sour, and the acidic, for these seem to be just about the distinct sorts of flavors. Therefore, what is capable of tasting is what is of these sorts in potency, and what can be tasted is what makes it be actively one of them.

Chapter 11 About the tangible and touch the account is the same, for if touch is not one sense but more than one, the tangible perceptible things would also have to be of more than one kind. But it is an impasse whether it is one or more than one sense, and what the sense organ that is perceptive of touch is, whether it is the flesh, and what corresponds to this in other animals, or whether it is not, but the flesh is the medium, while the first sense organ is something else inside it. For every sense seems to be of one pair of contraries,

422b 20

as sight is of white and black, hearing of high and deep pitch, and taste of bitter and sweet, but many pairs of contraries are inherent in what is tangible: hot/cold, dry/moist, hard/soft, and whatever others are of this sort. It is a certain resolution for this impasse at least, that in the other senses too there is more than one pair

422b 30

of contraries, for instance in sound not only highness and depth, but also loudness and quietness, smoothness and roughness of sound, and other things of that sort. And there are also other such distinctions about color. But it is not clear what the one thing is that underlies touch in the way that sound underlies hearing.

But whether the sense organ is inside, or is not, but

423a

is the flesh immediately, does not seem to be indicated by the perception's coming about at the same time as the being touched. For even as it is, if one were to stretch something around the flesh, making it a sort of membrane, *it* would similarly make the perception known immediately upon being touched, and yet it is clear that the sense organ is not in it (and if it were to grow into the flesh, the perception would go through it still more quickly); hence such a part of the body seems to act as though there were a natural circle of air around us, for then we would think that we perceived sound and color and smell by one organ, and that sight-

423a 10

sound-smell was a single sense. But as it is, since that through which the motions come is separate, it is clear that the sense organs mentioned are different. But in the case of touch this is now unclear, for it is impossible for the ensouled body to be made of air or water, since it has to be something solid; what is left is that it is a mixture of earth with air and water, as flesh and the things corresponding to it present themselves to be.

And so necessarily the body that is the medium for the organ of touch is grown upon it, and through it come perceptions that are of more than one kind. And touch with the tongue makes it clear that they are of more than one kind, since it perceives everything tangible with the same part that perceives flavor. So if the rest of the flesh also perceived flavor, it would seem that taste and touch were one and the same sense; as it is they seem to be two since they do not answer to each other in every way. 423a 20

But one might find an impasse, thus: if every body has depth, and this is its third kind of extension; and, when there is a body between two bodies, those two are not able to touch each other; and what is liquid or fluid cannot have being without a body, and must either be water or contain water; and things that touch each other in water, since their extremities are not dry, must have water between them, with which their surfaces are saturated; and if all this is true, it is impossible for one thing to touch another in water, and for the same reason even impossible in air. (For air has the same relation to the things in it that water has to the things in water, but this escapes our notice more, just as the animals in water, being wet, do not notice if they touch something wet.) So is the perception of all things similar, or is it different for different senses, just as now it seems that taste and touch perceive by being in contact, but the other senses from a distance? But this is not so, but we perceive even the hard and the soft through other things, just as we also perceive what can make a sound or a sight or a smell; but the latter are perceived from far away, the former from near at hand, and hence we are unaware of it, even 423a 30 423b though we do perceive them all through a medium, but in the former cases it escapes our notice. And yet, just as we were saying before, if we perceived all 423b 10 tangible things through a cloth membrane, it would go unnoticed that they were separated, we are in just the same situation now in water and in air; for now we seem to touch things which we do not perceive through a medium.

But tangible things do differ from visible and audible ones, because we perceive the latter kinds of thing when the medium acts upon us in some way, but the tangible ones not by the action of the medium but at the same time as the medium is acted upon, just as someone who is struck through a shield; for it is not that he is knocked by the shield's beating on him, but at the same time both he and the shield together are struck. And generally it seems that, as the air and water are in relation to sight, hearing, and smell, so too flesh and the tongue are related in just the same way to the sense 423b 20 organ that belongs to each of them. Neither here nor there would any perception happen if the sense organ itself were touched, for instance if someone were to put some white body on the surface of the eye. It should also be clear that what has the potency of the sense of touch is inside, for thus it would turn out to be exactly the same as in the case of the other senses; for when something is placed on the sense organ it is not perceived, but when something is placed on the flesh it is perceived, so that flesh is the medium of the sense of touch.

Now the tangible things are the distinctive attributes of body as body; by distinctive attributes I mean those that define the elements as hot, cold, dry, and moist, about which we have spoken before in the writings about the elements.[15] And the sense organ that has the potency of touch, in which what is called the sense of touch first inheres, is the part that is potentially of one of those attributes. For perceiving is a way of being acted upon, in which what acts makes another thing, which is potentially such as it, be of that attribute that the former has actively. For this reason, we do not perceive what is as hot or cold, or hard or soft, as we are, but what exceeds us, since the sense is a kind of mean between the contrary attributes in the things perceived. In virtue of this it discriminates the things perceived, for the mean has the discriminating power, since it comes to be either of two extremes in relation to the other. And just as the thing that is going to perceive white or black cannot be actively either one of them, but must be potentially both (and so too in the other cases), in the case of touch as well, the thing that perceives must be neither hot nor cold. And just as sight in a certain way is of both the visible and the invisible, and similarly with the rest of the perceptible objects, so too touch is of the tangible and the intangible. The intangible things are both those that have the distinctive attribute of the tangible ones to a very small degree, as air does, and those that have excessive degrees of them, so as to be destructive of the power to sense them. So each of the senses has been spoken of in outline.

423b 30

424a

424a 10

15. See *On Coming to Be and Passing Away*. Previous thinkers called earth, air, fire, and water the elements, and in Bk. I Aristotle was using their language. But he regards each of those four as a combination of the true elements, of hot or cold with moist (fluid) or dry. True elements could not change into one another, as the four simplest kinds of bodies do.

Chapter 12 But about all sense perception in general, it is necessary to grasp that the sense is receptive of the forms of perceptible things without their material, as wax is receptive of the design of a ring without the iron or gold, and takes up the golden or bronze design, but not as gold or bronze;[16] and similarly the sense of each thing is acted upon by the thing that has color or flavor or sound, but not in virtue of that by which each of those things is the kind of thing it is, but in virtue of that by which it has a certain attribute, and according to a ratio. The sense organ is the first thing that has the potency to be acted upon in that way, so the organ and the potency refer to the same thing, but the being of them is different; for the thing that does the sensing must be something extended, while its being a sense, or the being-perceptive of it, is certainly not of any size, but is a relatedness and potency of the part that has size.[17] These things also make clear why excesses of perceptible attributes sometimes destroy the sense organs (for if the motion produced is stronger than the sense organ can hold, the ratio—which is the sense—is done away with, just

424a 20

424a 30

---

16. This sentence has been the source of much controversy and confusion. Some of the difficulty can be avoided if one keeps in mind that, by form, Aristotle does not mean shape or appearance. In the Platonic dialogues, it is always emphasized that the *eidos* is not the mere look of a thing, but its invisible look, grasped by looking away from its visible attributes. That to which one must look, according to Aristotle, is the being-at-work of the thing. It is, then, no part of the analogy Aristotle intends with the wax impression, that the eyeball, say, becomes shaped and colored like an olive when it looks at one. What it takes on must be some sort of active condition.

17. The word relatedness here, and ratio just above and below, translate *logos*.

as tuning and pitch are when the strings of an instrument are struck too hard), and why plants do not ever perceive, even though they have a certain part of the potency of soul and are acted upon in some way by tangible things (since they become cold or hot)—the reason is that they have no mean condition, nor any source of such a kind as to receive the perceptible attributes, but they absorb them together with their material. 424b

One might be at an impasse whether something that cannot smell is acted upon in any way by a smell, or something that cannot see by a color, and similarly with the other senses. But if what is smelled is a smell, if it does anything at all the smell produces a perception of smell, so nothing that cannot smell is able to be acted upon by a smell (and the same argument applies also to the other senses), nor can anything that can perceive be acted upon by perceptible things except in the way in which each of them has the potency to perceive. The same thing is also clear from this: neither light and dark nor sound nor smell does anything to bodies, but that in which they are does; for instance, it is the air the thunder is in that splits the tree. The tangible attributes and flavored juices, though, do act on bodies, for if they do not, what is it by which soulless things would be acted upon and altered? Does that mean that the other kinds of perceptible things too would act upon unperceiving things? Or is it not every body that is able to be acted upon by smell and sound, while those that are acted upon are indeterminate and do not persist, for example air (which has a smell as though it had been acted upon in some way)? So what then is the smelling, over and above the being acted upon? Isn't smelling 424b 10 perceiving, whereas the air that has been acted upon immediately becomes something perceptible?[18]

---

18. From the time of the earliest commentaries and down to the most recent journal articles, most interpreters of *On the Soul* have tried to push it to one or the other of two extreme attitudes. On one type of reading, Aristotle is a materialist, who regards sense perception as a bodily event; on the other, he treats the perceiving soul as disconnected from and inexplicable by bodily events, as a "mind" or "consciousness" or "intentional" domain, dwelling in the body like a sailor in a boat. To take these as the only alternatives, and to incline toward one of them, reflects only the limitations of the thinking of the interpreter. Aristotle says in the *Physics* (194a 14–15) that no natural event happens either without material or as determined by material, and he carries this guiding principle through *On the Soul* all the way to the ultimate point at which it finds it own limit.

When Aristotle begins this chapter with the conclusion that in all perception the sense receives the form without the material, this is in no sense a "definition" (as the Loeb translator and others label it), but the first step in a long process by which he seeks to understand the potency that perceives the world. The air receives the form without the material from something that has a smell without thereby perceiving anything. At the other end of the spectrum, in us, no act of sense perception is isolated from our highest intellectual activity, by which we both discriminate what we perceive from and relate it to the thinghood of things. In air, the bodily event of reception of form is insufficient to produce perception; in our perception, the bodily event makes an indispensable contribution to the experience that goes beyond it.

Introductory Note to Book III

The word that has perhaps the widest range of meanings in *On the Soul* is *noein*, which refers to any and every kind of thinking. As used at 406b 25, it names the most rudimentary kind of discriminating, directed to sensory images alone, by which an animal with even the simplest powers of sensing and moving guides itself this way rather than that. As used in Book III, Chapter 5, it names an unchanging contemplative activity separate from all living things, that is neither born nor dies. Both these powers—the one subhuman, the other more than human—are also present in us, but the investigation of thinking in *On the Soul* is not restricted to the human soul. Our thinking is most characteristically reasoning things out, reckoning things together, thinking things through by steps, or deliberating about the best means to an end, but these activities are focused on in other works of Aristotle. Here the topic is the nature and possibility of thinking as something that must be built into the whole of things, not as a sidelight or "epiphenomenon" but as part of the working of the world. Book III, from beginning to end, from the combining of the senses, to the stillness and wholeness of contemplation, to the practical pursuit of an object of desire, investigates thinking. In its narrow use, *noein* means contemplation, and the power of which it is the action is *nous*, or the contemplative intellect. But since *noein* is also the comprehensive word for all thinking, translators are tempted to use the general word "mind" for that which thinks. The word "mind," though, is not precise and general but flabby and vague, and to the extent it carries meaning it brings in connotations from philosophers of the seventeenth century and later of an inert container of miscellaneous "ideas." Here, the word "mind" is used only once, in the translation of a quotation from Homer at 427a 26, where Aristotle criticizes the underlying quasi-bodily conception of thinking as crude and false. The potencies of the soul that are put to work in thinking are very many, even unlimited in number (432a 24, 433b 2), and Aristotle is interested in distinguishing and gaining clarity about some of the most important ones. The effect of his inquiry is not to blend and muddy the thinking power into one "mind" but to increase one's awareness of the variety of its workings. In fact, though they are not all topics of exploration, the following two dozen words for kinds of thinking are used throughout *On the Soul*:

ἀληθεύειν (*alētheuein*) to be right, to attain truth

ἀναμιμνήσκεσθαι (*anamimnēskesthai*) to recollect

ἀπορεῖν (*aporein*) to be at an impasse

βουλεύεσθαι (*bouleuesthai*) to deliberate

γιγνώσκειν (*gignōskein*) to discern

γνωρίζειν (*gnōrizein*) to recognize

διανοεῖσθαι (*dianoeisthai*) to think things through

δοξάζειν (*doxazein*) to suppose, to have an opinion

ἐπίστασθαι (*epistasthai*) to know

ζητεῖν (*zētein*) to inquire

θεωρεῖν (*theōrein*) to contemplate

θιγγάνειν (*thigganein*) to touch (the thing known)

κρίνειν (*krinein*) to judge, to distinguish

λογίζεσθαι (*logizesthai*) to reason

μανθάνειν (*manthanein*) to learn, to understand

μαντεύεσθαι (*manteuesthai*) to guess at, to divine

μνημονεύειν (*mnēmoneuein*) to remember

νοεῖν (*noein*) to think, to contemplate

οἴεσθαι (*oiesthai*) to suppose, to believe

συλλογίζεσθαι (*sullogizesthai*) to reckon together into a conclusion

ὑπολαμβάνειν (*hupolambanein*) to conceive, to suspect

φαντάζεσθαι (*phantazesthai*) to imagine

φρονεῖν (*phronein*) to have understanding

ψεύδεσθαι (*pseudesthai*) to be in error

Book III

Chapter 1 One might feel confident that there is no sense other than the five (by which I mean sight, hearing, smell, taste, and touch) from the following argument. For if we now have perception of everything of which the sense is touch (for all the attributes of the tangible, insofar as it is tangible, are perceptible to us by touch), and if, for some sense to be missing, we must also be missing some sense organ, but all things of which we have perception by being in contact with them are perceptible by means of touch, which we do happen to have, while those things which produce perception through a medium and not while we are in contact with them do so by means of the simple bodies—I mean, say, air and water—and are such that, when more than one sort of perceptible things in distinct classes from each other come through one medium, one who has that sort of sense organ is necessarily able to perceive both sorts (for instance, if the sense organ were made of air and air were the medium of both sound and color), while if one sort comes through more than one medium, as color comes through both air and water (since both are transparent), one who has either sort of sense organ alone will perceive what comes through both, and if sense organs are made of only two of the simple bodies, that is, of air and water (for the eyeball is made of water, the part that can hear is made of air, and the part that can smell can be made of either of these), while fire belongs either to none of them or in common to them all (for nothing that lacks heat is able to perceive) and earth either belongs to none of them or else is mixed in especially with the part that can perceive touch in particular, so

424b 22

424b 30

425a

that nothing would be left for a sense organ to be made of outside of water and air, and these some animals have even now; *therefore,* the animals that are not undeveloped or defective specimens have all the senses there are. (Even the mole obviously has eyes under its skin.) So if there is no other sort of body, nor any attribute which belongs to any of the bodies around us, no sense would be missing.[1]

425a 10

But neither is it possible for there to be a particular sense organ for the common attributes which we perceive as accessories by means of each of the senses, attributes such as motion, rest, shape, magnitude, and number; for we perceive all these—a magnitude by its motion (and thus also a shape, since the shape is a particular magnitude), or a thing at rest by its not moving, or a number by its absence of continuity—by means of the particular senses (for each sense has perception of one kind of thing). So it is clear that it is impossible for there to be a particular sense for any of these, say for motion, since this would just be the way we now perceive sweetness by means of sight; but this is because we happen to have perception of both of two things, by which means we recognize them when they concur together. And if that were not so we would not perceive these things at all, other than incidentally (as we perceive the son of Cleon, not perceiving that he is the son of Cleon but that he is pale-colored, which the son of Cleon incidentally happens to be). But of the things perceived as common we already have a common perception, which

425a 20

1. It is reported that Democritus claimed that gods, wise men, and lower animals have senses that the rest of us don't.

is not incidental, and so there is no particular perception of them, since then we would not perceive them at all, except in the way just described.

The senses perceive the things proper to each other incidentally, not by what each sense is in itself but by that which is one in them, whenever the perception occurs at the same time in the same object, for instance that bile is both bitter and yellow (for it certainly doesn't belong to either of the senses to say that one thing has both attributes). This is why there are mistakes, and if something is yellow it is supposed that it is bile. And one might inquire for what purpose we have more than one sense, and not just one. Is it so that the common things that accompany each sense, such as motion, size, and number, might be less able to escape notice? If there were only sight, and it perceived something white, the common attributes would escape our notice more, and would all seem to be the same, on account of their accompanying each other together with a color and a size. As it is, since the common attributes are also present in a different perceived thing, this makes it clear that each of them is something distinct.

425a 30

425b

425b 10

Chapter 2 But since we do perceive that we see and hear, it is necessary either to perceive by means of sight that one sees, or by means of some other sense. But in the latter case, the same sense would perceive both the seeing and the color that is its object, so either there would be two senses perceiving the same thing, or one that itself perceives itself. Also, if there were another sense to perceive sight, either this must go on to infinity, or there will be some sense that perceives itself, so one might as well make sight that way in the first place. But then there is an impasse, for if perceiving by sight is seeing, and what is seen is either color or what has color, then if one were to see the thing that sees, the thing that sees in the first place will have a color. It is clear then that perceiving by sight is not of only one sort, for even when we are not seeing, we discriminate the darkness from the light by means of sight, though not in the same way we discriminate colors. And also, it is in a way as though the thing that sees is colored, since the sense organ is in each case receptive of the perceptible attribute without the material; that is why, even when the perceptible things have gone away, there are still perceptions and images in the sense organs.

425 b20

And the being-at-work of the perceptible thing and of the sense that perceives it are one and the same, though the being of them is not the same—I mean, for instance, of the sound during the course of its being-at-work and the hearing during the course of its being-at-work, since it is possible for something having hearing not to be hearing, and something having sound is not always sounding—but when the thing capable of hearing is at work and the thing capable of making a sound is sounding, then the hearing in its being-at-work and the sound in its being-at-work come into being together, and one might call the former the hearing-activity and the latter the sounding-activity. So if a motion (both the acting and the being acted upon) is in the thing moved, then necessarily both the sound and the hearing in their being-at-work are in something that has being as a potency, since the being-at-work of the thing that acts and causes motion takes place in the thing that is acted upon; that is why a thing that causes motion is not necessarily in motion. So the being-at-work of what is such as to produce sound is sound or the sounding-activity, and the being-at-work of what is

425b 30

426a

such as to engage in hearing is hearing or the hearing-activity, for hearing is twofold, and sound is twofold.[2]

The same account applies also to the other senses and perceptible things. For just as both the acting and the being acted upon are in the thing being acted upon and not in the thing acting upon it, so too the being-at-work of the perceptible thing and that of the thing capable of perceiving it are in the thing that can perceive. In some cases both have names, such as sounding and hearing, but in other cases one of them is nameless, for the being-at-work of sight is called seeing, but that of color is nameless, and that of the sense of taste is tasting, but that of flavor is nameless. And since the being-at-work of the perceptible thing and the perceiving sense are one, though the being of each is different, it is necessary that hearing and sound meant in that way, or flavor and tasting, shut off or endure together, and so too with the other senses, but this is not necessarily so if these things are meant as potencies. But the earlier thinkers about nature did not give a good account of this, since they supposed that there is neither white nor black without seeing, nor flavor without tasting. In one way what they said was right, but in another way it was not right, since the sense and the thing perceived are meant in two ways, either as in potency or as at work, and in the latter meaning what they said applies, but in the former meaning it does not apply. But they spoke in a simple way about things that are not meant in a simple way. Now if sound is a kind of consonance, and there

426a 10

426a 20

2. Hearing is twofold because the activity of the sense organ is also the action of the bell, while sound is twofold because it is either mere sounding or the complete act of sounding-being-heard.

is a way in which sound and hearing are one, and consonance is a ratio, then hearing too is necessarily a kind of ratio. For this reason each kind of excess, the high as well as the low pitch, shuts off hearing, and similarly an excess among flavors shuts off taste, among colors the too-bright or too-dark shuts off sight, and among smells, a strong smell, sweet or bitter, shuts off that sense, inasmuch as the sense is a particular ratio. This is also why something purified and unmixed is pleasant when it is brought into a ratio, such as something sour or sweet or salty, for that is when they are pleasing, and generally what is mixed is more harmonious than what is sharp or flat, and what is warmed or cooled is more pleasant to touch; sense perception is ratio, and what is excessive undoes or destroys ratio.

426a 30

426b

So each sense is directed to the perceptible thing that is its object, is present in the sense organ insofar as it is a sense organ, and discriminates the distinctions that belong to the perceptible thing that is its object, sight, say, discriminating white and black, but taste sweet and bitter, and this is the same way with the other senses. But since we also distinguish white from sweet, and each perceptible thing from every other, then there is also something by which we perceive that they are different. And this is necessarily perception, since they are perceptible things. This also makes it clear that the flesh is not the final sense organ, since then the thing that distinguishes perceptible attributes would have to be in contact with them in order for it to distinguish them. And neither is it possible for separate senses to judge that sweet is different from white, but it is necessary that both be apparent to some one thing—otherwise, even if I perceived one of them and you perceived the other, it could be evident that they were different from each

426b 10

426b 20

other, but it is necessary that some one thing say that they are different,[3] for sweet *is* different from white and therefore the same thing says so, and what it says, so too it both grasps by thinking and perceives—so it is clear that it is impossible for separate senses to distinguish what is in fact distinguished.

That they cannot be distinguished even in separate times is clear from the following. For just as it is the same thing that says that good and bad are different, so too *when* it says that the one is different it also says so of the other (and the when is not incidental—I mean the way it is when I *say* now that they are different rather than that they are different now—but it both says so now and says that it is so now); therefore they are distinguished at the same time, so that it is an indivisible act in an indivisible time. But surely it is impossible that at one time the same thing be moved in opposite motions, insofar as it is undivided and is considered in an undivided time. For if something is sweet, it moves the sensing or the grasping by thinking in one way, but something bitter moves it in an opposite way, and something white in a still different way. So then is the thing that distinguishes them undivided and indivisible in number but divisible in its being?[4] Then

426b 30

427a

3. William James made this sort of argument famous in the nineteenth century: "Take a sentence of a dozen words, and take twelve men and tell to each one word. Then stand them in a row or jam them in a bunch, and let each think of his word as intently as he will; nowhere will there be a consciousness of the whole sentence." (*Psychology, the Briefer Course*, University of Notre Dame Press, 1985, p. 66.)

4. An example of this is the road from Athens to Thebes, which is the same in number with the road from Thebes to Athens, while its being the one and the other are two different things.

there is a sense in which something divided perceives divided things, and a sense in which it does so as undivided, since it is divided in being, but undivided in place and number. Or is this impossible? For something that is the same and undivided is opposite things in potency, but not in its being, but is divided by being put to work, and it is impossible at the same time to be white and black, so that it is also impossible to be acted upon by the forms of white and black at the same time, if that is what the sensing of them or the grasping

427a 10

them by thinking is. But like what some people say about the point, insofar as it is both one and two, in this way it too is divided. So insofar as it is undivided, the thing that distinguishes is one thing and distinguishes things at one time, but insofar as it has it in it to be divided, it uses the same boundary-mark[5] as double at one time. So insofar as it uses a boundary as double, and distinguishes two separate things, it acts in a way dividedly, but insofar as it acts by means of one thing, it acts as one and at one time.

So let the source by which we say that an animal has the power of perception be marked out in this way.

Chapter 3 But since people define the soul most of

---

5. To translate *sēmeion* as "impression" would push the meaning in the direction of later materialist thinkers; to translate it as "symbol" would push it toward later philosophers of consciousness. Aristotle uses it for the design on a ring, but primarily for any kind of sign or indication that something is the case. It becomes Euclid's word for the mathematical point, by derivation from its original reference to a boundary-mark, but Aristotle prefers the more vivid word *stigmē*, derived from an original reference to a puncture. In *On Memory and Recollection*, 450b 15–27, Aristotle speaks of a "trace" (*tupos*) left by sense-perception, which the soul can look at, as an image, or look beyond, as a sign.

all by two distinct things, by motion with respect to place and by thinking, understanding, and perceiving, while thinking and understanding seem as though they are some sort of perceiving (for in both of these ways the soul discriminates and recognizes something about beings), and the ancients even say that understanding and perceiving are the same—as Empedocles has said "wisdom grows for humans as a result of what is present around them," and elsewhere "from this a changed understanding is constantly becoming present to them," and Homer's "such is the mind" means the same thing as these,[6] for they all assume that thinking is something bodily like perceiving, and that perceiving and understanding are of like by like, as we described in the chapters at the beginning (and yet they ought to have spoken at the same time about making mistakes as well, for this is more native to living things and the soul goes on for more time in this condition, and thus it would necessarily follow either, as some say, that everything that appears is true, or that a mistake is contact with what is unlike, since that is opposite to recognizing like by like, though it seems that the same mistake, or the same knowledge, concerns opposite things)—nevertheless it is clear that perceiving and understanding are not the same, since all animals share in the former, but few in the latter.

427a 20

427a 30

427b

And neither is thinking the same as perceiving, for in thinking there is what is right and what is not right,

---

6. *Odyssey* XVIII 130–137: "Nothing feebler than a human being does the earth nourish, of all the things that breathe and crawl on the earth...for the mind of earth-dwelling humans is such as what the father of men and gods sends each day." The Empedocles fragments are 106 and 108 in the Diels numbering.

427b 10 right thinking being understanding and knowing and true opinion, and the opposites of these not being right; for sense perception when directed at its proper objects is always truthful, and is present in all animals, but it is possible to think things through falsely, and this is present in no animal in which there is not also speech. For imagination is different both from perceiving and from thinking things through, and does not come about without perception, and without it there is no conceiving that something is the case. And it is clear that imagination is not the same activity as conceiving that something is so,[7] for the former experience is available to us whenever we want it (for it is possible to make something appear before the eyes in the 427b 20 way people do who make images to fit things into a memory-assisting scheme) but to form an opinion is not up to us, since it has to be either true or false. Also, when we have the opinion that something is terrifying or frightening we immediately feel the corresponding feeling, and similarly if we think it is something that inspires confidence, but with the imagination we are in the same condition as if we were beholding terrifying or confidence-inspiring things in a painting. There are different ways of conceiving that something is the case—knowledge, opinion, understanding, and their opposites—but let the account of their differences be given elsewhere.

As for thinking, since it is different from perceiving, 427b 30 and since it seems that one sort of thinking is imagination and another sort of thinking is conceiving that

7. Descartes gives an unforgettable example of this difference: Conceive of an equiangular plane figure with 1000 equal sides; now imagine it.

something is the case, one ought to speak about the latter sort after having thoroughly distinguished what pertains to imagination. Now if imagination is that by which we speak of some image as becoming present to us, rather than anything we might call imagination in a metaphorical way, is it some one among those potencies or active states by which we discriminate something and are either right or wrong? Of this sort are perception, opinion, knowledge, and the contemplative intellect. But that imagination is not perception is clear from the following arguments. For perception is either a potency or a being-at-work, such as sight and seeing, but images appear also in some way when neither of these conditions is present, such as the ones in dreams. Next, perception is present in every animal, while imagination is not, but if they were the same in their activity, then in all the animals imagination would be capable of being present, though it seems not to be, since it is in an ant or a bee, for example, but not in a larval worm.[8] Next, perceptions are always true, but most imaginings turn out to be false. Next, when we are engaged accurately with some perceived thing we do not say, for example, that we imagine this is a human being, but we say this instead when we do not perceive plainly whether that is true or false. And, as we were saying

428a

428a 10

8. The manuscripts get the last part of this sentence wrong; the translation follows Ross's text, which reconstructs the original from evidence in ancient Greek commentaries. Linda Wiener, an entomologist, points out to me that ants and bees, with their complex nests and hives, coordinated nest defense strategies, and ability to gather and store food for the winter, plainly act as though aware of things distant in time or place. The anomalous, and completely different, instance of larval worms, which are born before all their sense organs have developed, is addressed in Chapter 11 below.

before, visual images appear even to those whose eyes are shut.

But certainly imagination would not be any of the things that are always truthful, such as knowledge or the contemplative intellect, since there is also imagining that is false. It remains, then, to see if it is opinion, since both true and false opinions occur. But belief goes along with opinion (since it does not seem possible for someone to have an opinion who does not believe it), but belief is present in none of the other animals, while imagination is present in many of them. Moreover, while belief accompanies opinion, having been persuaded accompanies belief, and speech accompanies persuasion, and while imagination is present in some of the other animals, speech is not.

428a 20

It is clear, then, that imagination could not be opinion along with sense perception, nor by way of sense perception, nor an intertwining of opinion and sense perception, both for the reasons above, and because, if that is what it is, there would then be no opinion of anything else except that of which there is also perception. I mean that imagination would be an intertwining of the opinion and the perception that something is white, but not of the opinion that it is good with the perception that it is white. In that case the imagining would be the having an opinion of the very thing one perceives, and not one incidentally related to it. But it is possible even for there to be a false appearance of things about which, at the same time, there is a true conception; for example, the sun appears to be a foot wide, but one believes it to be bigger than the inhabited world. It follows either that one has lost sight of one's own true opinion, which one held, while the thing remained the same and one

428a 30

428b

neither forgot it nor was persuaded otherwise, or, if one still holds it, then one must hold the same opinion to be true and false. But a true opinion could become false only if one had failed to notice that the thing had undergone a change. Therefore, imagination is neither one of these things, nor anything made out of both.

But since it is possible when one thing is moved for another thing to be moved by it, while imagination seems to be some sort of motion and not to occur without perception, but in beings that perceive and about things of which there is perception, and since it is possible for a motion to come about as a result of the being-at-work of sense perception, and necessary for it to be similar to the perception, then this motion would be neither possible without perception nor present in beings that do not perceive, and the one having it would both do and have done to it many things resulting from this motion, which could be either true or false. This last point follows because, while sense perception of its proper objects is true or has the least possible falsehood, there is in the second place the perception that those things that are incidental to the ones perceived are in fact incidental to them, and here already it is possible to be completely mistaken, not mistaken that something is white, but that the white thing is this or that other thing. And in the third place there is perception of the common attributes that accompany the things incidentally perceived, to which the things properly perceived belong (I mean, for instance, motion or size), about which most of all it is possible to be deceived as a result of sense perception. And the motion that comes about from the activity of perception, stemming from these three ways of perceiving, will be different in each case, the first sort

428b 10

428b 20

being truthful while the perception is present, while the others could be false whether it is present or absent, and especially when the thing perceived is far away.

428b 30 429a If, then, it is nothing other than imagination that has the attributes mentioned (and this is what was being claimed), imagination would be a motion coming about as a result of the being-at-work of sense perception, and corresponding to it.[9] And since sight is the primary sense, imagination (*phantasia*) has even taken its name from light (*phaos*), because without light it is impossible to see. And because imaginings remain within and are similar to perceptions, many animals act in accord with them, some, the beasts, because of not having intelligence, but others, humans, because their intelligence is sometimes clouded by passion, disease, or sleep. So about imagination, let this much be said about what it is and the cause through which it comes about.

429a 10 **Chapter 4** About the part of the soul by which the soul knows and understands, whether it is a separate part, or not separate the way a magnitude is but in its meaning, one must consider what distinguishing characteristic it has, and how thinking ever comes about. If thinking works the same way perceiving does, it would either be some way of being acted upon by the intelligible thing, or something else of that sort. Therefore it must be without attributes but receptive of the form

9. Imagination is not only derivative from sense perception but derived as a motion from a being-at-work. A being-at-work is exactly what it is through time, and could not consist of motions within a sense organ. Those motions lead up to an active state of perception in the soul, which produces in turn the consequent motions of imagination. See also the note to 409a 16.

and in potency not to be the form but to be such as it is; and it must be similar so that as the power of perception is to the perceptible things, so is the intellect to the intelligible things. Therefore necessarily, since it thinks all things, it is unmixed, just as Anaxagoras says, in order to master them, that is, in order to know them (since anything alien that appeared in it besides what it thinks would hinder it and block its activity); and so intellect has no nature at all other than this, that it is a potency. Therefore the aspect of the soul that is called intellect (and I mean by intellect that by which the soul thinks things through and conceives that something is the case) is not actively any of the things that *are* until it thinks. This is why it is not reasonable that it be mixed with the body, since it would come to be of a certain sort, either cold or warm, and there would be an organ for it, as there is for the perceptive potency, though in fact there is none. And it is well said that the soul is a place of forms, except that this is not the whole soul but the thinking soul, and it is not the forms in its being-at-work-staying-itself, but in potency. 429a 20

The absence of attributes is not alike in the perceptive and thinking potencies; this is clear in its application to the sense organs and perception. For the sense is unable to perceive anything from an excessive perceptible thing, neither any sound from loud sounds, nor to see or smell anything from strong colors and odors, but when the intellect thinks something exceedingly intelligible it is not less able to think the lesser things but even more able, since the perceptive potency is not present without a body, but the potency to think is separate from body. And when the intellect has come to be each intelligible thing, as the knower is said to do when he is a knower in the active sense (and this happens when 429a 30 429b he is able to put his knowing to work on his own), the intellect is even then in a sense those objects in potency, but not in the same way it was before it learned and discovered them, and it is then able to think itself.

429b 10 Now since a magnitude is different from being a magnitude, and water is different from being water (and so too in many other cases, though not in all, since in some cases the two are the same), being flesh is distinguished either by a different potency from the one that distinguishes flesh, or by the same one in a different relation. For flesh is not present without material, but like a snub nose, it is this in that.[10] So it is by the perceptive potency that one distinguishes hot and cold, and the other things of which flesh is a certain ratio, but it is by a different potency that one distinguishes the being-flesh, either separate from the first or else with the two having the relation a bent line has to itself straightened out.[11] Among the things that have being in abstraction, straightness is in its way just like snubness, since it is combined with continuity; but 429b 20 what it is for it to be, if being straight is different from what is straight, is something else—let it be twoness.[12]

10. Snubness is an example Aristotle often uses of something that has its underlying material in its very meaning. A mathematical curve might resemble its shape, but could not replace it in thought.

11. This might refer to seeing a stick partly immersed in water. It is also interesting to compare it with Plato's *Phaedo*, 74A–75B, where Socrates describes our seeing unequal sticks as though they were straining to attain the equality we think in its purity; sense perception stimulates a "recollection" that gets hold of the same objects in a different way. Steven Werlin has pointed out to me that the Greek leaves undetermined which of these potencies gets things straight.

12. The straight line has two dimensions; two points determine it. In some way twoness seems to govern what it is. Aristotle uses the

Therefore one distinguishes it by a different potency, or by one in a different relation. So in general, in whatever way things are separate from their material, so too are the potencies that have to do with intellect separate from one another.[13]

But one might find it an impasse, if the intellect is simple and without attributes and has nothing in common with anything, as Anaxagoras says, how it could think, if thinking is a way of being acted upon (for it seems to be by virtue of something common that is present in both that one thing acts and another is acted upon), and also whether the intellect is itself an intelligible thing. For either there would be intellect in everything else, if it is not by virtue of something else that it is itself intelligible, but what is intelligible is something one in kind, or else there would be something mixed in it, which makes it intelligible like other things. As for a thing's being acted upon in virtue of something common, the distinction was made earlier, that the intellect *is* in a certain way the intelligible things in potency, but is actively none of them before it thinks them; it is in potency in the same way a tablet is, when nothing written is present in it actively—this is exactly

429b 30

430a

phrase "things that have being in abstraction" only of mathematical things. See 431b 12-17.

13. This rich and difficult paragraph implies that intellect pervades all human experience. What Aristotle has repeatedly called incidental perception, the recognition that this pale white shape is the son of Diares or Cleon (probably students in front of him), seems to be the same act as distinguishing flesh or water. Thus the things that we perceive are already organized in accordance with something intelligible, and one of the things the intellect thinks is the perceptible thing in its wholeness. The same perceptible form that acts incidentally on the various sense organs acts directly on intellect, but is not the only sort of form that the intellect takes on.

what happens with the intellect. And it is itself intelligible in the same way its intelligible objects are, for in the case of things without material what thinks and what is thought are the same thing, for contemplative knowing and what is known in that way are the same thing (and one must consider the reason why this sort of thinking is not always happening); but among things having material, each of them is potentially something intelligible, so that there is no intellect present in them (since intellect is a potency to be such things without their material), but there is present in them something intelligible.

430a 10 **Chapter 5** But since in all nature one thing is the material for each kind (this is what is in potency all the particular things of that kind), but it is something else that is the causal and productive thing by which all of them are formed, as is the case with an art in relation to its material, it is necessary in the soul too that these distinct aspects be present; the one sort is intellect by becoming all things, the other sort by forming all things, in the way an active condition such as light does, for in a certain way light too makes the colors that are in potency be at work as colors. This sort of intellect is separate, as well as being without attributes and unmixed, since it is by its thinghood a being-at-work, for what acts is always distinguished in stature above what is acted upon, as a governing source is above the
430a 20 material it works on. Knowledge, in its being-at-work, is the same as the thing it knows, and while knowledge in potency comes first in time in any one knower, in the whole of things it does not take precedence even in time. This does not mean that at one time it thinks but at another time it does not think, but when separated it

is just exactly what it is, and this alone is deathless and everlasting (though we have no memory, because this sort of intellect is not acted upon, while the sort that is acted upon is destructible), and without this nothing thinks.[14]

Chapter 6 The thinking of indivisible things is one of those acts in which falsehood is not possible, and where there is falsehood as well as truth there is already some kind of compounding of intelligible things as though they were one—just as Empedocles says "upon the earth, foreheads of many kinds sprouted up without necks"[15] and then were put together by friendship, so too are these separate intelligible things put together, such as incommensurability and the diagonal—and if

430a 30

430b

14. This brief chapter is the source of a massive amount of commentary and of fierce disagreement. Much of it centers around the question whether the topic of the chapter is an activity within the human soul or something outside us. In a similar way, one might ask whether the motionless first mover deduced in the *Physics* is a cause within or outside nature, or whether the contemplative life described in the *Nicomachean Ethics* is still a human life or somehow beyond that. The ambiguity in all these instances is inevitable, dealing with the place where each inquiry reaches its limit and points beyond itself. Human thinking, natural motion, and a life that absorbs the whole human capacity for happiness all depend on something that differs from them in kind. That something is, in each of these cases, the same being, and it is found in the *Metaphysics* to be that on which all being depends, a self-subsistent activity of thinking that thinks itself. Our thinking, on occasion, may merge into this activity and become identical with it, despite the fact that, on its side, its action is continuous and unfailing. But its thinking must be always at work in or upon us, to make possible our humblest acts of perceiving and knowing.

15. This passage, which continues,"and arms wandered unattached, bereft of shoulders, and eyes strayed about alone, lacking brows," is fragment 57 in Diels's numbering.

the thinking is of things that have been or are going to be, then one puts them together while additionally thinking the time. For falsehood is always in an act of putting things together, for even in denying that white is white one puts together not-white with white; or it is also possible to describe all these as acts of dividing. At any rate, not only is it possible for it to be true or false that Cleon is pale-skinned, but also that he was or will be pale. What makes each thing be one is the intellect.

But since an undivided thing can be of two kinds, either incapable of division or actively undivided, nothing prevents one from thinking an undivided thing when one thinks it for some duration (since it is actively undivided), nor from thinking it in an undivided time, for a time is divided and undivided in the same way as its duration. So it is not possible to say what one thought in each half of the time, since there are no halves when it has not been divided, other than potentially. But by thinking separately during each of the halves, one also divides the time along with it, and then it is as if there were separate durations; but if one thinks in a way that is made out of the two halves, one also thinks in a time that applies to both.

430b 10

But what is indivisible not in amount but in kind, one thinks in an indivisible time and by means of something indivisible in the soul, but incidentally, and not in the way those things by means of which one thinks and the time in which one thinks are divisible, but in the way that they are indivisible, for even in these there is something indivisible, though perhaps not separate, that makes the time and its duration one. And this is similarly in every continuous thing, a time as well as a length. A point, and every division, and what is indivisible in that way, is evident in the way a

430b 20

deprivation is. And a similar account applies to other things, such as how one recognizes what is bad or black, since one recognizes them in a way by means of their opposites. But to do this the knower must potentially be those opposites and they must potentially be in him; if there is any knower to which nothing is opposite, it knows itself and is a being-at-work and separate.

Every act of saying something about something, and likewise of denying, is also either true or false, but this is not so with every act of intellect, but thinking what something is, in the sense of what it keeps on being in order to be at all, is true, and is not one thing attached to another.[16] But in the same way that the seeing of something proper to sight is true, but seeing whether the white thing is a human being or not is not always true, the same thing holds also with thinking the things that are without material. 430b 30

Chapter 7 Knowledge, in its being-at-work, is the same as the thing it knows, and while knowledge in potency comes first in time in any one knower, in the whole of things it does not take precedence even in time, for all things that come into being have their being from something that is at-work-staying-itself. And it is 431a

---

16. Recognizing something as a horse, say, or an olive tree, is not an act of thinking a collection of attributes together as a sum. Empedocles, quoted earlier in this chapter is consistent; his argument about plants, referred to in Book II, Chapter 4 above, denied that anything holds the roots and branches together as one organized whole, so body parts must be separate and independent formations, attached externally. Aristotle is also consistent. The form that organizes the plant or animal also works upon the perceiving or thinking soul that recognizes that plant or animal. It is ultimately the active intellect of the preceding chapter that forms and produces these wholes.

obvious that the perceptible thing makes the perceptive thing be at-work from being in potency, for the perceiving thing is not [merely] acted upon nor is it altered. Hence this is a different kind of event from a motion, since motion is a being-at-work of something incomplete, while being-at-work in the simple sense, that of something complete, is different again. And perceiving is similar to simple declaring and to thinking contemplatively; but when the thing perceived is pleasant or painful, the one perceiving pursues or flees it, as though affirming or denying.

431a 10 Being pleased or pained is the being-active of the mean state in the perceptive part, in relation to the good or bad as such, and the fleeing and the desiring, in their being-at-work, are the same thing, nor are the desiring part and the fleeing part different from each other or from the perceiving part, though the being of them is different. And for the soul that thinks things through, imaginings are present in the way perceptible things are, and when it asserts or denies that something is good or bad it flees or pursues; for this reason the soul never thinks without an image.[17] It is the same way that the air acts on the eyeball in a certain way, and this acts on something else, and the same with hearing, while the last thing acted upon is one thing with a single mean

---

17. This last clause occurs in a context. It is not the thinking soul simply which is being discussed, but the soul that thinks things through and concludes about pursuing or avoiding something; at 433a 14–15 below, this is understood as the same capacity as the contemplative intellect, but distinguished in activity by being turned to a practical use. Imagination allows practical thinking to go beyond present perceptions and consider distant possibilities and novel combinations of things. The following paragraph gives an example of thinking something through outside a practical context.

condition, though the being of it is of more than one kind.

It was said before what it is by which the soul distinguishes how sweet and hot differ, but one ought to speak of it again in the following way. It is one thing, but in the same way a boundary is, and since those attributes are analogous and belong to one thing, each of them bears the same relation to its opposite that the other bears to its opposite. So why should the question of how one distinguishes attributes of different kinds be any different from how one distinguishes opposites such as white and black? So as A, white, is to B, black, let C be to D; therefore, alternately, as A is to C, so is B to D. Now if C and A were to belong to one thing, they, and also D and B, would in that way hold of something that is one and the same, while not being the same,[18] and similarly with D and B. The argument would be the same if A were sweet and B were white. 431a 20 431b

Now the thinking potency grasps in thought the forms that are present in things imagined, and since what is to be pursued or fled from is marked out for it in those imaginings, even apart from sense perception,

18. The upshot of the argument is that sweet and white, or any such pair, while not being directly comparable, are alike in being possible attributes of the same thing. Thus the alternate ratios make some sense here, as they would not if A were a length and C an area. In the light of the claim made in the preceding paragraph, it is interesting to ask whether, when the soul replaces the objects of its thought with letters as place holders for some steps of its reasoning, it is thinking without an image. The A, B, C, and D are useful just because they put aside any imagining of sweet or hot or white things, and permit us to concentrate on general relations of indefinite attributes. But it might be argued that the letter A itself is an image, a minimal residue of imagining that we cannot do without while thinking.

it is moved when it applies itself to imagined things. For instance, perceiving that a signal light is a fire, and observing by what is common to the senses that it is moving, one recognizes that it is an enemy; but sometimes, by means of the imaginings and thoughts in the soul, just as if one were seeing, one reasons out and plans what is going to happen in response to what is present. And when the soul declares, as it would in the case of perceiving, something pleasant or painful,

431b 10 here in this case too one flees or pursues it, and so in all matters of action. But also apart from action, the true and the false are present in the same classes of things as the good and bad, but they differ in that the former are present simply, the latter in relation to someone. The so-called abstractions are just the sort of thing one gets if one were to think of, say, a snub nose, not as snub but separately as a concave shape, if one should think it without the flesh in which the concave shape is. In this way one thinks the mathematical things, which are not separate from material, as though they were separate, whenever one thinks them. But in all cases the intellect, in its being-it-work, *is* the things it thinks. Whether it is possible for something to think any separated thing without itself being separate from extended magnitudes, or not, must be considered in the next section.

431b 20 **Chapter 8** And now, bringing together what has been said about the soul under one main point, let us say again that the soul is in a certain way all beings, for beings are either perceptible or intelligible, while knowledge in a certain way is the things it knows, and perception is the things it perceives; but one needs to inquire in what way this is so. Now knowledge and perception are divided up into the things they are concerned with, and there is in potency know-ledge or perception to be divided into the things that are in potency, and knowledge or perception at-work-staying-itself that is divided into the things that are at-work-staying-themselves; so what the perceiving and knowing capacities of the soul are in potency are the same things that are either known or perceived. This has to be either those things themselves or their forms; it is certainly not themselves, since a stone is not present in the soul, but its form is. Thus the soul is like a hand, for the hand is a tool of tools, while the intellect is a form of forms[19] and sense perception is a form of perceptible things. 432a

But since—as it seems—there can be no item of experience apart from the extended magnitudes which are the separate perceptible things, the intelligible things are present in the perceptible forms, not only the so-called abstractions but all the active conditions and passive attributes of the sensible things. And on account of this, one who perceived nothing would not be able to learn or be acquainted with anything either, and, whenever one were to contemplate, it would be necessary at the same time to behold some image. For the things imagined are just like the things perceived, except without material. And imagination is different from affirmation and denial, since what is true or false is an intertwining of intelligible things. So how do the 432a 10

19. The genitive in these two phrases has as varied a meaning as our preposition "of." The hand is a tool that uses tools, or shapes itself to fit various tools, or stands before and opens the way to many tools, or is pre-eminent among tools.

uncombined intelligible things differ from being images? But in fact these are not images either, but are not present without images.[20]

Chapter 9 But since the soul is distinguished by two potencies of living things, that of discriminating, which is the work of reasoning and of sense perception, and that of causing motion with respect to place, let the preceding things that concern perception and intellect be the explication of them, but about that which produces motion, one must examine whatever it is that belongs

432a 20

to the soul that does this, whether it is some one part of it that is separate either as a magnitude or in its articulation, or else the whole soul, and if it is some part, whether it is some special one besides the ones usually

---

20. This paragraph is full of pitfalls for the unwary. The assumption it begins with cannot mean there is no independent thing at all apart from extended bodies. The unextended first cause of motion deduced in Book VIII of the *Physics*, which is identified in Book XII of the *Metaphysics* as the source of the wholeness of every genuine being, and is found again here in Chapter 5 above to be a precondition for all thinking, is such a thing. The way that separate intellect thinks the forms is not the question here. We encounter the forms only in the things we experience, the *pragmata*. For us, learning begins always with perception and our highest acts of contemplation are still linked with images. Can we infer anything at all about the thinking of the active and productive intellect? If its thinking is what forms the things we perceive, its contemplation too would always be linked with images, not in itself but as the source of the sources of all such images. Our thinking is always rooted in and dependent upon images for its content; the thinking of the separate intellect might be imageless in its own right, but such that it always made possible an image by which souls could ascend to that highest kind of thinking. In *On Memory and Recollection* (449b 31), Aristotle says without qualification that there is no thinking without an image, but one must consider, for any particular kind of thinking, whether that image is an ingredient in, a precondition for, a consequence of, or a potential counterpart to what is thought.

spoken of and already discussed, or some one of these. There is immediately an impasse, both as to the sense in which one ought to speak of parts of the soul, and how many there are. In a certain way they seem unlimited in number, and not just the ones some people speak of, dividing it up into reasoning, spirited, and desiring parts, or as others do, into what has reason and the irrational part. For in accord with the distinctions through which they separate these parts, the soul also obviously has other parts farther apart than these are, and which have been spoken about here: the nutritive part, which belongs both to the plants and to all animals, and the perceptive part, which one could not easily place either as irrational or as having reason, and also the imaginative part, which in its being is different from all the rest, though whether it *is* the same or different from any of them[21] is a major impasse, if one is going to set down the parts of the soul as separate, and in addition to these the appetitive part which would seem to be different from them all both in its articulation and in its potency. And it is surely absurd to tear the soul apart in this way, since wishing ends up in the reasoning part, but desire and spiritedness in the irrational part; and if the soul is three things, there will be appetite in each of them.

432a 30

432b

And especially there is the question the account is now set upon: what is it that makes a living thing move with respect to place? For since motion in the sense of growing and wasting away is present in them

21. See 429b 10–22. The imagination might be the part of the soul that discriminates flesh and water, while turned differently the same part might discriminate intelligible things, such as what flesh is. See the distinction made below at 433b 29 and 434a 5–7.

432b 10 all, the thing that belongs to them all, the potency to beget and to use nourishment, would seem to be what produces the motion. What concerns breathing in and out, and sleep and waking, must be considered later,[22] since these too have major impasses, but one must examine what concerns motion with respect to place, and what causes the animal to move in the sense of passing through distance.

It is clear that this is not the nutritive potency, since this sort of motion is always for the sake of something, and goes along with imagination and desire, for nothing moves without desiring or fleeing something, except by force; also, plants too would be able to move, and would have some part instrumental for this sort of motion. Similarly, it is not the perceptive potency, 432b 20 for there are many among the animals that have perception but are in one place and motionless throughout life;[23] but if nature neither does anything in vain, nor leaves out anything necessary except in living things that are mutilated or incompletely developed, while such animals are fully developed and are not mutilated (an indication is that they are able to beget offspring and have a peak and decline of life)—then they too would have parts that were instruments of passage.

And it is certainly not the reasoning part, or intellect in the way that word is used, that causes motion, for the contemplative intellect does not contemplate anything that has to do with action, and says nothing about what is to be fled from or pursued, while motion always belongs to a being that is fleeing something or pursuing

22. There are separate brief treatises on each of these topics.

23. See the note to 410b 19–20.

something; and even when the thinking part does contemplate something of that sort, it still does not urge that one flee or pursue it, such as often when it thinks about something frightening (or pleasant) but does not urge one to fear it, even though the heart is moved (or, if it is pleasant, some other part). And even when the intellect enjoins and the reasoning part declares that something is to be fled or pursued one does not necessarily move, but acts instead in accordance with desire, as does one without self-restraint. And generally we see that the one who has medical knowledge does not necessarily heal anyone, since it is something else that governs one's doing anything in accordance with the knowledge and not the knowledge itself. But neither is it desire that governs this sort of motion, since self-restrained people, even when they desire and yearn for something, do not necessarily do those things for which they have the desire, but follow the intellect. 432b 30 433a

Chapter 10 But it is obvious that these two things cause motion, desire and/or intellect, if one includes imagination as an activity of intellect, since many people follow their imaginings contrary to what they know, and in the other animals there is no intellectual or reasoning activity, except imagination. Therefore both of these are such as to cause motion with respect to place, intellect and desire, but this is intellect that reasons for the sake of something and is concerned with action, which differs from the contemplative intellect by its end. Every desire is for the sake of something, since that for which the desire is, is the starting point for the intellect concerned with action, and its last step is the starting point of the action. So it is reasonable that there seem to be two things causing the motion, 433a 10 desire and practical thinking, since the thing desired causes motion, and on account of this, thinking causes 433a 20 motion, because it is the desired thing that starts it. Imagination too, when it causes motion, does not do so without desire. So it is one thing that causes motion, the potency of desire; for if the two, intellect and desire, caused the motion, they would do so as a result of some form common to them, but in fact the intellect obviously does not cause motion without desire (for wishing is desiring, and whenever one is set in motion in accordance with reasoning, one is also set in motion in accordance with wishing), while desire causes motion even contrary to reasoning, since a passionate impulse is a kind of desire. And while every act of the contemplative intellect is right, desire and imagination can be both right and not right. Hence it is always the desired thing that causes motion, but this is either the good or the apparent good, and not every apparent good, but the good as contained in action. But what 433a 30 concerns action is what admits of being otherwise.

It is clear then that this sort of potency of the soul, 433b the one called appetite, causes motion. And for those who divide up the soul into parts, if they divide and separate them in accordance with their potencies, a great number of them come about—the nutritive part, the perceptive, the intellective, the deliberative, and now the appetitive—for these differ more from each other than do the desiring and spirited parts. But since desires come to be opposite to one another, which happens whenever reason and impulses are opposed, and comes about in beings that have perception of time (for the intellect urges one to resist impulses on account of the future, while the impulse urges one to resist reason on account of what is immediate, since what is immediately pleasant appears to be both simply pleasant and simply good, on account of not looking to the future), then while the thing that causes motion would be one in kind, the desiring part as desiring—or first of all the thing desired, since it causes motion without being in motion, by being thought or imagined—there come to be a number of things that cause motion. And since there are three kinds of thing, of which one is the thing causing motion, the second that by which it causes motion, and the third is the thing moved, while the thing causing motion is of two sorts, the one motionless and the other in motion as well as causing motion, the motionless cause of motion is the good sought by action, while the cause of motion that is in motion is the desiring part of the soul (for the thing moved is moved by desiring something, and the desire when at work is itself a kind of motion), the thing moved is the animal, and the instrument by which desire causes motion is already part of the body, for this reason one must study what concerns animal motion among the acts performed by the body and soul in common. For now, to speak in the manner of a summary, what causes motion as an instrument of it is at a place where the same thing is a beginning and an end, like a hinge, for there the convex and concave parts are the end and beginning (which is why one part is still and the other in motion), being different in meaning but not separated in extension. For all things are moved by a process of pushing and pulling, for which reason it is necessary, as in a wheel, for something to stay still and for the motion to start from there.[24] In general then, as was said,

433b 10

433b 20

24. The part that ends up moving, the arm or leg or wing or fin,

it is as having the potency of desire that an animal is capable of moving itself, but the potency of desire is not present without imagination, while all imagination 433b 30 is either rational or sensory, and of the latter the other animals have a share.

Chapter 11 But one must consider also what it is that causes motion in those animals that are incom- 434a pletely developed, in which perception is present only by touch, and whether it is possible or not for them to have imagination and desire.[25] Now pleasure and pain are obviously present in them, and if these are, then desire necessarily is present as well. But in what way could imagination be in them? Is it that, just as they move in an unspecific way, imagination too is present in them, but is in them in an unspecific form? So a sensory imagination, as was said, is present in the rest of the animals, while there is a deliberative imagination in those that can reason (for whether one will act this way or that way is already a job for reasoning, and has to be measured by one criterion, since one is looking for the greater good, and thus is able to make one 434a 10 thing out of a number of images). This is the reason the other animals do not seem to have opinion, because they do not have opinion that comes from reckoning things together. Hence desire does not have the power

---

must have a still beginning place to push against, a hinge-like joint, even though that part is incidentally moved by being carried along with the animal. The rim of a wheel similarly must be fixed against the hub at the center if the axle is to turn it. The heart of the animal is thought of as its center, from which all motion first radiates. These things are discussed in the *Parts of Animals* III, 4 and the whole of the *Motion of Animals*.

25. These are the larvae mentioned at 428a 11.

of deliberating, but at one time this desire wins out and knocks away that one, and at another time that one wins out and knocks away this one, like a ball, when there is a lack of self-restraint; but by nature the higher desire is more governing and causes the motion—so that there are already three ways of being moved. But the power of knowing is not moved, but stays constant. But since one sort of assumption or proposition is universal and the other sort particular (for the one says that a certain sort of person ought to act in a certain sort of way, the other that this particular action is of that sort and I am that sort of person), it is surely the latter sort of opinion that causes motion, and not the universal, or else it is the two together, but with the universal, not the particular, remaining more constant. 434a 20

Chapter 12 So everything that lives and has a soul at all necessarily has the nutritive soul, from birth and up to death; for what has been born must have growth, a peak of life, and a decline, and these are impossible without nourishment. Therefore it is necessary that the nutritive potency be in everything that grows and decays, but perception is not necessary in all living things, for those of which the body is simple are unable to have it, as are those which are not receptive of forms without material; but an animal needs to have perception, and without this it is not possible to be an animal, if nature does nothing in vain. For everything that is by nature is present for the sake of something, or else is something that necessarily accompanies the things that are for the sake of something. And if any body were such as to pass through distance, but did not have perception, it would be destroyed and would not reach the end which is the work of its nature. (For 434a 30 434b how would it feed itself? Stationary living things have food from that out of which they have been born, but it is not possible for a body to have a soul and an intellect that can distinguish things but not have perception, if it is not stationary and has been generated—and even if it were ungenerated—for why would it not have perception? Only if that were better for the soul or for the body, but in fact it is so for neither, since the one will not think any more nor the other be anything more on account of that.) Therefore no body that is not stationary has a soul without perception.

But surely if it does have perception, a body must be either simple or mixed, but it cannot be simple, since it would not have the sense of touch, and it is necessary for it to have that.[26] This is clear from the following argument. For since an animal is an ensouled body, and every body is tangible, it is necessary that the body of the animal be capable of perceiving by touch as well, if the animal is going to preserve itself. For the other senses, such as smell, sight, and hearing, perceive through the medium of something else, so a body that is in contact with something, if it were not to have perception of it, would be unable to flee some things and grasp others, and if that were so it would be impossible for the animal to preserve itself. And this is why taste acts like a kind of touch, since it is perceptive of food, and food is a tangible body. But sound, color, and smell do not nourish, nor do they produce either growth or wasting away, so taste must be a kind of touch since it is perceptive of what is tangible and nutritive.

434b 10

434b 20

---

26. This is explained below at the beginning of Chapter 13.

So touch and taste are necessary senses for an animal, and it is clear that without touch it is impossible for the animal to be, but the other senses are for the sake of well-being, and they already are not in any random class of animals, though in some, such as the sort that pass through distance, they must be present, for if these are going to preserve themselves, they must perceive not only when they are in contact with something but also from a distance. And this could happen if the animal could be perceptive through a medium, when the medium is acted upon and moved by the perceptible thing, and it by the medium. For just as that which moves something with respect to place causes part of the changing up to a point, and that which pushes acts on something else so that it pushes as well, and the motion is through an intermediate thing, and the first mover pushes without being pushed, the last is pushed only, without pushing anything, and the intermediate thing both pushes and is pushed, and the intermediate things are many, it is the same way in the case of qualitative alteration, except that something could alter while staying in the same place. For example, if one were to dip something in wax, the wax would be moved up to the point that the thing was dipped; stone would not be moved by that means at all, but water would be set in motion to a great distance, and air is moved and acted upon and acts up to the greatest distance, if it stays intact and is one. That is why, concerning reflection, rather than saying that sight goes out and bounces back, it is better to say that the air is acted upon by shape and color for as far as it is one, and since up against a smooth body the air is one, this moves the sight in turn, just as if a design in wax had spread through to the other side of the body.

434b 30

435a

435a 10

Chapter 13 It is clear that it is not possible for the body of an animal to be simple, by which I mean, say, purely fire or purely air. For without the sense of touch, it would not be able to have any other sense (since every ensouled body is such as to perceive by touch, as was said); the other elements excluding earth could have become sense organs, but all of them cause the sense to perceive by way of something else and through a medium, while the sense of touch is by touching what it perceives, and hence has that name. And yet the other sense organs too perceive by touch, but through something else, while touch seems to perceive through itself alone. Therefore, the body of an animal could not be made of any of the elements other than earth, nor could it be purely earth. For the sense of touch acts as a mean condition among all the tangible attributes, and its sense organ is receptive not only of the distinctive attributes of earth, but also of hot as well as cold and of all the other tangible attributes. It is for this reason that we do not perceive by means of our bones or hair or parts such as these, because they are made of earth, and that is why plants have no sense perception at all, because they are made of earth, while without touch no other sense can possibly be present, and the sense organ of touch is not made of earth nor of any other one of the elements alone.

435a 20

435b

It is clear then that, if they were deprived of this sense alone, animals would necessarily die; for neither is it possible to have this sense and not be an animal, nor is it necessary in order to be an animal to have any other sense except this one. This is why the other perceptible attributes, such as color, sound, and smell, do not, in their excesses, destroy the animal, but only its sense organs (except incidentally, as in case a push

435b 10

or a blow came together with a sound); by the action of visible things or of a smell other things are set in motion which destroy by means of touch (and a flavored juice, insofar as it belongs to it at the same time incidentally to be tangible, destroys by means of that), but the excess of any tangible attribute, such as heat or cold or hardness, does away with the animal. For the excess of any perceptible attribute does away with its sense organ, so also the excess of what is tangible does away with the sense of touch, and by this the animal is determined, since it has been shown that without touch it is impossible for the animal to be. On account of this, the excess of tangible attributes destroys not only the sense organ, but also the animal, because this is the only sense it needs to have. The other senses the animal has, as was said, not for the sake of being but for the sake of well-being—for example sight, since it lives in air or water, or generally in a transparent medium, in order that it may see, and taste, in order that it may perceive, by means of what is pleasant and painful, what is in food, and desire it and be moved, and hearing in order that something may be signified to it, and a tongue in order that it may signify something to another. 435b 20

On Memory and Recollection

Introductory Note to
On Memory and Recollection

The following text is the second in a collection of eight works generally known as the *Parva Naturalia*, or short writings pertaining to nature, though Aristotle describes their topics as activities "common to the soul and the body" (*On Sense-Perception and Perceptible Things*, 436a 8–9). The whole collection presupposes the things said in *On the Soul*, but the text included here is the one that is closest to the inquiry into the soul itself, and least connected with opinions about the workings of the body. It is an especially valuable extension of a number of things said in *On the Soul*. Only in *On Memory and Recollection* does Aristotle say without qualification that there is no thinking without an image. Only here does he make it plain that he understands the imagination to be the primary or common perceptive power over and above the five senses, a single potency that perceives time as well as the common and incidental perceptible things dealt with in *On the Soul*. Only here does he describe the double power of imagination to behold the traces left by sense perception on their own as pictures of a sort within the soul, but also as likenesses or signs of the original perceived things, so that the image (*phantasma*) is both an appearance in itself and something pointing beyond itself. This in turn allows Aristotle to be more explicit here about his claim that the other animals have intelligence, though they lack reason. Finally, it is here that Aristotle discusses the connection of recollection with, on the one side, the inherent interrelations of mathematical things, and, on the other, the incidental and external relations later writers called the "association of ideas."

Memory *(mnēmē)*, the active holding of an image as a likeness of a past perception, a power shared by most animals, is the topic of Chapter 1. Recollection *(anamnēsis)*, the human power of deliberately taking up again something known or perceived before, is explored in Chapter 2.

On Memory and Recollection

Chapter 1 About memory or remembering, one must say what it is, and through what cause it comes about, and of which of the parts of the soul this experience and recollecting are attributes. For it is not the same people who are good at remembering and good at recollecting, but for the most part those who are slow are better at remembering while those who are quick and learn easily are better at recollecting. First, then, one must consider of what sort the things are that can be remembered, since one is often deceived about this. For it is not possible to remember the future, but it is something about which there is opinion and expectation (and there might even be a certain sort of knowledge pertaining to expectation, as some people say about skill in prophecy); nor is there memory of the present, but rather perception, for by this we become acquainted with neither the future nor the past, but the present only. But memory is of the past. 449b 4 449b 10

That which is present, when it is present, such as this white thing here when one is seeing it, no one would claim to be remembering, nor that which one contemplates, when one happens to be contemplating and thinking it (for instance that the angles of a triangle are equal to two right angles),[1] but one claims only to perceive the former and know the latter; but whenever one has the knowledge or the perception without the acts, in that way one remembers that one understood or contemplated the latter, or heard or saw, or something of that sort in the former case. For whenever one is 449b 20

1. The words in parentheses follow the word "remembers" in line 20, but some editors suggest they belong here.

actively engaged in remembering, one says in that way in one's soul that one heard or perceived or thought this before. Memory, then, is neither perception nor understanding, but is an active condition or passive state of one of these, whenever time has passed. Of the now, in the now, there is no memory, as was said, but there is perception of what is present, expectation of what is in the future, and memory of what is past; hence, every memory is involved with time. And so, among the animals, only those that perceive time remember, 449b 30 and they do so by means of that by which they perceive it.

And following what was said before about imagination in the writings on the soul, it is not possible even 450a to think without an image. For the same condition goes along with thinking which goes with drawing a diagram, since there, while making no use of the triangle's being of a definite size, still we draw it definite in size; and in the same way, one who is thinking, even if one is thinking of something that is not a quantity, sets a quantity before one's eyes, though one does not think it *as* a quantity, but if the nature of it is among things that have a quantity, but an indefinite one, one sets out a definite quantity, but thinks it just as a quantity. What the reason is that it is impossible to think anything apart from something continuously extended, or to think beings that are not in time apart from time, is another story, but it is necessary to become acquainted with 450a 10 magnitude and motion by means of that by which one is also aware of time, so it is clear that the acquaintance with these is by means of the primary power of perception, while memory, even memory of intelligible things, is not without an image, and the image is an attribute of the common perceiving power, so that memory would belong incidentally to the intellect, but in its own right it belongs to the perceptive potency.[2] For this reason memory is present also in some other animals, and not only in human beings and those animals that have opinion and intelligence.[3] But if it were one of the intellectual parts [of the soul] it would be absent from many of the other animals, and perhaps not present in any mortal being, since even now it is not in all animals because they do not all have perception of time; for always, whenever one is at work with one's memory, just as we said before, remembering that one saw or heard or learned this particular thing, one perceives in addition that one did this earlier, and earlier and later are in time. It is clear, then, to which of the powers of the soul memory belongs, that it is the very one to which imagination also belongs, and the things remembered in their own right are those of which there is imagination, while as many things as are not apart from imagination are remembered incidentally. 450a 20

---

2. The common perceiving power (*hē koinē aisthēsis*) is not just the power that directly perceives the common perceptible things—motion, rest, number, shape, magnitude, and time—but is a common potency that goes along with each of the senses (*On Sleep and Being Awake* 455a 15–16), incidentally perceives intelligible wholes such as the son of Cleon, is aware that we are perceiving, discriminates among the objects of the different senses, and perceives images when the perceptible thing has gone away (*On the Soul* 425a 14–b 25). It is thus the primary perceptive power as distinct from the particular powers of sight, touch, etc. The third to last clause in this sentence is moved to a more logical position by editors, from its manuscript position following the words "aware of time."

3. These are not just the "higher" vertebrates, since Aristotle calls bees more intelligent than many of them (*Parts of Animals* 648a 6–7).

But someone might be at an impasse about how, when the thing one is concerned with is absent, but the experience of it is present, one remembers something that is not present. For it is clear that one must conceive the sort of thing that comes about in the soul—and in the part of the body that contains it—as a result of sense perception as something like a picture, the active holding of which we assert to be memory. For the motion that comes about traces in something like an outline of the thing perceived, in the same way people mark designs into things with rings. This is why, in people who are in vigorous motion on account of passion or their time of life, memory does not come about, just as if the motion and its impression fell upon flowing water; in others, on account of being worn down like old walls and because of hardness in the receptive part that is acted upon, the outline does not get into it. For these reasons, both the very young and the old are lacking in memory, for the former are in flux because they are growing, the latter because they are decaying. And similarly, neither the very quick nor the very slow display good memories, since the former are more fluid than is needed and the latter are more hardened; the image does not remain in the soul in the former, and it does not attach itself to the latter.

450a 30

450b

450b 10

But if what goes on in the case of memory is of this sort, does one remember this experience or the one from which it came about? For if it is this one, we would not remember any of the things that are absent, but if it is that earlier one, how, while perceiving this later one, do we remember the absent thing that we are not perceiving? And if there is something like a tracing or a picture in us, why would the perception of this be a memory of something else and not of this very thing?

For the one who is actively remembering is beholding this experience and perceiving it. In what way then will one remember the thing that is not present? One should then be able also to see and to hear something that is not present. Or is there a sense in which this is possible and does happen? For example, the picture drawn on a tablet is both a picture and a likeness,[4] and one and the same thing is both of these, although what it is to be these two things is not the same, and it is possible to behold it both as a picture and as a likeness; so too one ought to conceive of the image that is in us as being itself something in its own right, and as being *of* something else. Insofar, then, as it is something in its own right, it is a thing beheld or an image, but insofar as it is of something else, it is a certain kind of likeness or reminder. 450b 20

And so, whenever the motion is at work that belongs to it insofar as it is something in its own right, if the soul perceives it by this motion, a certain sort of thought or image seems to come before it; but if, insofar as it is of something else, the soul beholds it in the same way that one beholds what is in a picture as a likeness and, when one has not been seeing Coriscus, 450b 30

4. A "likeness" (*eikōn*) means not any and every thing that resembles something else (*homoiōma*), but something derivative that imitates or represents an "original"; it is found in Plato's *Sophist* (240B 12–C 2) to be a paradoxical sort of thing that is what it is by in some way being what it is not. The word "image" is reserved in this translation for *phantasma*, the likeness that is present in the soul. The soul's power of imagination (*phantasia*), in the strict sense that Aristotle is discussing, is thus an instance of a more encompassing power, called *eikasia* in Books VI and VII of Plato's *Republic*, of understanding things as pointing beyond themselves to some original. See Jacob Klein, *A Commentary on Plato's Meno* (University of North Carolina Press, 1965), pp. 112–115.

as Coriscus, and the experience of this beholding then, and when one beholds it as a painted picture, are different, in the soul too it comes about in the one way only as a thought, but in the other way, because, as in the case of the painting, it is a likeness, it becomes a reminder. And for this reason sometimes we do not know, when motions of this sort come about in our souls after earlier perceiving, whether they follow as a result of having been perceived, and we are in doubt whether it is a memory or not; but sometimes it so happens that we consider and recollect that we heard or saw something before. And this happens whenever, beholding the image as itself, one changes and beholds it as being of something else. And the opposite also occurs; for instance it happened to Antipheron of Oreus[5] and to others who were out of their senses. For they spoke of their imaginings as things that had happened and that they were remembering. This comes about whenever anyone beholds what is not a likeness as a likeness. But exercises preserve memory by reminding, and this is nothing else but frequently beholding images as likenesses and not in their own right.

451a

451a 10

What memory, or remembering, is, then, has been said, that it is an active holding of an image as a likeness of that of which it is an image, and it has been said to which of the parts in us it belongs, that it belongs to the primary perceptive power, by which we also perceive time.

---

5. This may the same person described in Aristotle's *Meteorology* (373b 4–10), who saw illusory images reflected back from the air around him.

Chapter 2 It remains to speak about recollecting. First, then, one ought to set down as starting points as many of the things in the exploratory writings[6] as are true. For recollection is neither the taking up again of memory nor the taking up of it. For when one first learns[7] or experiences something, one does not take up any memory again (since none has gone before) nor take one up from the start, for when the active condition or passive state has come in, then there is memory, so that it does not come in along with the experience when that is coming in. Besides, from the first, when the experience has come in to the indivisible extremities, it is already present in the one experiencing it, and so is the knowledge, if one ought to call this active condition or passive state knowledge (though nothing prevents us from incidentally also remembering some of the things we know), but to remember is not possible in its own right until time has passed. For one remembers now what one saw or experienced before; one does not remember now what one experienced just now. Also, it is clear that it is possible to remember things not by recollecting them now, but by perceiving or experiencing them from the beginning, but when one takes up again what one had before, the knowledge or perception or anything

451a 20

451a 30

451b

6. There is an ancient list of Aristotle's works that refers to three of these, but nothing else is known of them.

7. Aristotle's account of what happens when we first learn things in the deepest sense may be found in an unexpected place: *Physics*, Book VII, Chapter 3, 247b 1–6. Readers who may regret the prosaic treatment of recollection Aristotle gives here will find there that he has a metaphor of his own for learning that is as powerful as Plato's.

of which we said memory to be the active holding, this, at this time, is the recollecting of one of the things mentioned, but it is involved with remembering and memory follows it. And it is not even simply this, but there is a sense in which it is, and another sense in which it is not, for it is possible for the same person to learn or to discover the same thing twice. It is necessary, then, for recollecting to differ from this, and a source must be present within beyond that by which one learns in order for one to recollect. 451b 10

Recollections result because this or that motion naturally comes about after this or that other one; if this is by necessity, it is clear that whenever one is set in motion in that way, one will be set in motion also in this way, but if it is not by necessity but from habit, one will be set in motion this way for the most part. And it turns out that some motions are habituated when one is set in motion once, more than others are when one is set in motion often; for this reason we remember some things that we have seen once better than others we have seen often. When we recollect something, then, we set in motion one of the earlier motions, until we are set in motion with that one after which the thing recollected follows habitually. And this is why we hunt through what is in sequence, starting to think from what is present or from something else that is similar or opposite or neighboring to what is sought. 451b 20 By this means the recollection comes about, since the motions belonging to these things are either the same as, or simultaneous with, what is set in motion with that one, or contain a part of it such that the remainder is small.

So people search in this way, and even when they do

not search, this is the way they will recollect, whenever that comes about after another motion, but for the most part, that happens when other motions of the kinds we spoke of have come about. There is no need to consider how we remember things that are far removed, but only things that are nearby, since it is clear that the way is the same—I am speaking of things in a sequence that one has not searched for in advance nor recollected. For the motions follow one another by a habit, this one after that one, and so whenever one wants to recollect, one will do this: one will seek to get hold of a beginning of motion after which that thing sought would be. Hence recollections come about quickest and best from a beginning, for as the things one is concerned with stand toward one another in the sequence, so too do the motions. And all things are easily remembered that have some ordering, such as mathematical things have; the rest are remembered badly and with difficulty. And in this respect recollecting differs from learning things again, because one will be able in some way on one's own to be moved along to what follows the beginning. Whenever one is not able to do this except by the help of another, one no longer remembers. And often, one is unable to recollect something right away, but having searched for it, one is able and discovers it. This happens to someone who sets many things in motion until one sets a motion going that is of the sort from which the thing one is concerned with will follow. For remembering is the presence within of the power that sets the motion going, and this in such a way as to be set in motion out of oneself and those motions that one contains, as was said. But it is necessary to get hold of a starting point, which is why people sometimes seem to

451b 30

452a

452a 10

be recollecting from things from places[8]; the reason is that they go quickly from one thing to another, such as from milk to white, from white to mist, and from this to rainy, from which one remembers the autumn, when this is the time one is looking for.

In general, the middle of all things is like a starting point, for if one does not remember before, one will remember when one comes to this, or else not from any other. For example, if one thinks of things labeled 452a 20 ABCDEFGHI, if one does not remember at I, one will remember at E, since from there one is able to move in both directions to either D or F; if one is not seeking either of these, one will remember when one goes to C, if one is looking for A or B, or if not, when one goes to G, and so in all cases. And the reason for sometimes remembering, but sometimes not, from the same place is that it is possible to move in more than one direction from the same starting point, for instance from C to B or to D. So if one is moving along an old path,[9] one moves in the more accustomed direction, for habit by then is just like nature. Hence, what we often think of, we quickly recollect, for just as by nature 452a 30 this thing brings that thing with it, so too by being-at-work, and a thing that happens often produces a

---

8. In the *Topics* (163b 29–31), Aristotle says that someone skilled at memory remembers things by setting out *topoi*, presumably any contexts that include the things sought. This clause seems to mean that our spontaneous ways of getting obliquely at something forgotten resemble the devices used in methodical memory training.

9. The letters in lines 19–26 are badly scrambled in the manuscripts. The translation follows a reconstruction by Ross that is simple and likely enough. But Ross then needlessly turns the old path into a long lapse of time.

nature. And since, just as among things that are by nature, things happen also that are contrary to nature and from chance, this is still more so among things that are by habit, in which nature is not present in equal strength, so that one is moved sometimes in this direction and sometimes in another, both for other reasons and whenever something distracts one away from here to somewhere over there; and for this reason, when there is a need to remember a name, if we know one something like it, we make a slip of the tongue into that one. Recollecting, then, happens this way. 452b

And the most important thing one needs to recognize is the time, either by measure or in an indefinite way. And let there be something by which one distinguishes the time as more or less;[10] it is reasonable that this happens in the same way as with extended magnitudes. For one thinks of things that are big or far off not by having one's thinking stretch out to there, as some people say sight does[11] (for one would think of them in the same way even if they were not present), but by a proportional motion; for there are similar figures and motions in one's thinking. In what way, then, when one thinks of bigger things, will what one thinks in respect to those things be different from what it is when one thinks of smaller ones? Now all inner things are smaller, and in proportion, and perhaps, just as it is possible with figures to take another proportional one 452b 10

10. This is the common or primary perceiving power; see 450a 9–14 and the note there.

11. Empedocles and Plato's character Timaeus speak of light going out from the eye as from a lantern. See *On Sense Perception and Perceptible Things*, 437b 10–438a 5.

inside the figure itself, so too is it possible with time intervals. For example, then, if one is moved along AB, BE, that produces CD, for AC and CD are proportional [to AB and BE]. But why does it produce CD rather than FG? Or is it because AC is to AB as H is to I? So one moves along these in the same times. But if one wants to think of FG, one thinks in the same way of BE, but instead of H and I one thinks of J and K, since these are to one another as FA is to BA.[12]

452b 20

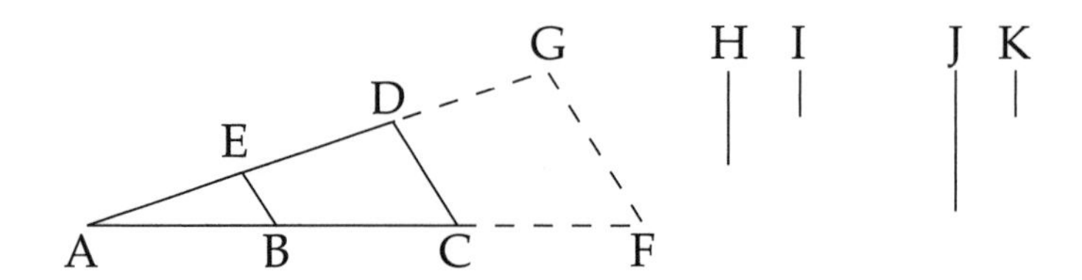

Thus, whenever the motion belonging to the thing one is concerned with unfolds concurrently with the movement belonging to the time, one is then actively

12. There is an incorrect diagram with some of the manuscripts that has the inner triangle centered in the outer similar figure, rather than sharing a vertex with it. The appropriate diagram needs to be of the same kind as that with Proposition VI, 12 of Euclid's *Elements*. The isolated lines stand for the times of events, while the sides of the triangles stand for motions in the imagination. These lines are not meant to be representational images of any content in the imagination, but play the same role the lines do in the "universal mathematics" of Book V of Euclid's *Elements*. That is, since they are magnitudes, their ratios and proportions can be the same as those of any other kind of magnitudes. Aristotle identifies the time of a motion as a magnitude in the *Physics*, 220b 24–32. The triangles have no special significance, but are just the simplest way to display a fourth proportional; a perception of the time interval to the thing recollected falls out in the imagination if the motions there keep the scale of the original events. The letters in this passage, like those in 452a, are scrambled in the manuscripts; they and the accompanying drawing are given here as reconstructed by Ross.

engaged with one's memory, but if one supposes this without doing so, one supposes that one remembers, for nothing prevents one from being mistaken in some way, and seeming to remember when one is not remembering. But it is impossible to remember without noticing it, if one is actively engaged with one's memory and not merely supposing so, for this very thing is the remembering.[13] But if the motion belonging to the thing one is concerned with comes about apart from that belonging to the time, or the latter apart from the former, one does not remember. But the motion belonging to the time is of two sorts, for sometimes one does not remember it by a measure, for instance that one did this or that two days ago, but sometimes also with a measure, but even without a measure one does remember. And people commonly say that they remember something, though they do not know when it was, whenever they do not distinguish the when of it by a measure of its amount. Now that it is not the same people who are good at remembering and good at recollecting was said in the previous chapter. But recollecting differs from remembering not only with respect to the time, but because, while many of the other animals have a share in remembering, none of the animals, insofar, one might say, as we know, except the human being, shares in recollecting. And the cause of this is that recollecting is a certain sort of reasoning; for the one who recollects reasons out that one saw or heard or had some such experience before, and this is a certain sort of inquiry. And this

452b 30

453a

453a 10

13. See 450a 29–30 and 451a 14–16. To be actively engaged with one's memory (*energein tē mnēmē*) is an active holding of an image as a likeness (*hexis tou phantasmou hōs eikonos*), and this active holding is what we would call attention or awareness.

belongs by nature only to those in whom a power of deliberation is also present, since deliberating is also a certain sort of reasoning.[14]

A sign that the experience is in some respect bodily, and that recollection is a search for an image within some part of the body, is that it greatly disturbs some people when they cannot recollect something even when they concentrate their thinking strongly, yet when they are no longer trying to recollect they do so nonetheless, especially impulsive people, for imaginings produce the greatest amount of motion in 453a 20 these people. And the reason why recollecting is not in one's own power is that, just as it is no longer in the power of those who throw something to make it stop, so too the one who is recollecting and hunting sets some part of the body in motion, in which the experience takes place. And the people in whom this disturbance is greatest are those in whom moisture happens to be present in the region of the part that is involved in perception,[15] since it does not easily

---

14. See *On the Soul*, 434a 5–15.

15. According to Aristotle, this part is not the brain but the heart. See especially *Parts of Animals*, 656a 23–31, and *On Youth and Old Age*, 469a 5–23. This explains his reference below to pressure on the organ of perception in people with heavy upper bodies. In the lecture described in the bibliographic note, Linda Wiener gives several reasons why Aristotle resisted the opinion, common even in his time, that perceiving and thinking depend on the brain. (1) In the chick embryo, the heart visibly develops first. (2) In all animals, sensing is present only in parts supplied with blood. (3) The heart is central, with blood flowing to it from all the outer parts that have sense perception. (4) Invertebrate animals, some of which are clearly intelligent, visibly have hearts or analogous organs, but not brains. (5) Animals that die naturally often have something wrong with the heart, while animals that are slaughtered often have tumors or lesions in the brain, but not in the heart.

stop moving until the thing sought has comes back and the motion has gone straight past it. This is why anger and fear, when something sets them in motion, do not settle down even when one opposes these in turn by counter-motions, but keep moving in the same direction against the resistance. And the experience is like that with names or tunes or phrases, when any of them have come to be much on the lips, for they keep coming back again to those who have stopped singing or saying them and do not want to. 453a 30

And those who have the upper body larger and have dwarf-like builds are worse at remembering than those with opposite builds, on account of having a lot of weight on the part involved in perception, and from the start the motions are not able to maintain themselves in them, but break up, nor are they easily able to stay on a straight course in recollecting. The exceptionally young and the very old are bad at remembering on account of the motion, for the latter are in decay while the former are in rapid growth; also, in the case of the children, they are of a dwarf-like build until well along in age. 453b

So about memory and remembering, what their nature is and by what part of the soul animals remember, and about recollecting, what it is and how it comes about and through what causes, have been said. 453b 10

Glossary

Aristotle's principal theoretical works, the *Physics*, the *Metaphysics*, and *On the Soul*, depend upon and illuminate one another in countless ways. This glossary contains the most important words and phrases that run through those works, and is based upon those passages in which Aristotle explains and clarifies his own usage.

The first section of this glossary lists, alphabetically, the Greek words that are discussed below, with the English translations that have, for the most part, been used for them here. In many cases, the translation chosen is followed in parentheses by the standard translation that is *not* used here. Some of these rejected translations, such as *perplexity* or *formula*, convey false impressions by wrong emphases; others, such as *habit* or *speculation*, are cognate to good Latin translations of the Greek, but have entirely different meanings in English; still others, such as *substance* or *induction*, are mistakes made long ago that have hardened into authority.

The second section of the glossary discusses all these words at some length, with reference to many of Aristotle's writings. The glossary thus supplements the introduction, and provides a way for you to orient yourself to Aristotle's primary vocabulary as a whole.

I. Greek Glossary

αἴσθησις . . . sense perception (*not* sensation)

αἰτία . . . cause

ἀλλοίωσις . . . alteration

ἀνάγειν . . . lead back (*not* reduce)

ἀντίφασις . . . contradictory

ἀπορία . . . impasse (*not* perplexity, difficulty)

ἀρετή ... virtue

ἀριθμός ... number

ἁρμονία ... harmony

ἀρχή ... source (*not* principle)

αὐτόματον ... chance

ἀφαίρεσις ... abstraction

γένος ... genus

διάνοια ... thinking things through

δύναμις ... potency (*not* potentiality)

εἶδος ... form

ἐναντίον ... contrary

ἐνέργεια ... being-at-work (*not* actuality)

ἐντελέχεια ... being-at-work-staying-itself (*not* actuality)

ἕξις ... active state

ἐπαγωγή ... example (*not* induction)

ἠρεμία ... rest

θεωρία ... contemplation (*not* speculation)

ἰδέα ... form

καθόλου ... universal

κατὰ συμβεβηκός ... incidental (*not* accidental)

κίνησις ... motion

κινοῦν ... mover (*not* efficient cause)

λόγος ... articulation (*not* formula)

μεταβολή ... *See* motion

μορφή ... form

νοεῖν . . . thinking

νοῦς . . . intellect (*not* mind)

νῦν . . . now (*not* moment)

ὁμωνυμία . . . ambiguity (*not* equivocation)

οὐσία . . . thinghood (*not* substance)

ποιόν . . . of-this-kind (*not* quality)

ποσόν . . . so much (*not* quantity)

πρῶτη, πρῶτον . . . primary

πρῶτη φιλοσοφία . . . first philosophy

στάσις . . . rest

στέρησις . . . deprivation (*not* privation)

στοιχεία . . . elements

τὰ μετὰ τὰ φυσικά . . . *See* first philosophy.

τέλος . . . end

τέχνη . . . art

τί ἐστι . . . *See* what it is for something to be

τί ἦν εἶναι . . . what it is for something to be (*not* essence)

τόδε τι . . . this (*not* this somewhat)

τόπος . . . place

τύχη . . . chance

ὕλη . . . material (*not* matter)

ὑποκείμενον . . . underlying thing

φαντασία . . . imagination

φύσις . . . nature

χωριστόν . . . separate

II. English Glossary

This is a slightly revised version of the glossary that appears with the translations of the *Physics* and the *Metaphysics*, based upon those passages in which Aristotle explains and clarifies his own usage. Bekker page numbers from 184 to 267 refer to the *Physics;* those from 402 to 435 refer to *On the Soul*; those from 980 to 1093 are in the *Metaphysics.*

abstraction (**ἀφαίρεσις** *aphairesis*) The act by which mathematical things, and they alone, are artificially produced by taking away in thought the perceptible attributes of perceptible things (1061a 29–1061b 4). Within mathematics, this is the ordinary word for subtraction. It is never used by Aristotle to apply to the way general ideas arise out of sensible particulars, as Thomas Aquinas and others claim. Its special philosophic sense is not Aristotle's invention; as often as not he speaks of "so-called abstractions." *Aphairesis*, which is the ordinary word for subtraction, is used by Aristotle in its philosophic sense rarely, only in reference to the origin of mathematical ideas, and not always then; in the *Physics* he says instead that mathematicians separate what is not itself separate (193b 31–35).

active state (**ἕξις** *hexis*) Any condition that a thing has by its own effort of holding on in a certain way. Examples are knowledge and all virtues or excellences, including those of the body such as health. Of four general kinds of qualities described in *Categories* Book VIII, these are the most stable. In the *Nicomachean Ethics*, the notion of a *hexis* is central to the inquiry. The mistranslation "habit" causes endless confusion, while the most common translations, "state" and "condition," lose almost all the meaning the word carries.

alteration (**ἀλλοίωσις** *alloiōsis*) Change of quality or sort, dependent upon but not reducible to change of place. One of the four main kinds of motion. Some things that we would consider qualities are "present in" the thinghood of a being, making it what it is, rather than attributes of it; change of any of them would be change of thinghood, rather than alteration of a persisting being (226a 27–29). The acquisition of virtue is just such a change of thinghood, not an alteration but the completion of the coming-into-being of a human being, just as putting on the roof completes the coming-into-being of a house (246a 17–246b 3). For a different reason, learning is not an alteration of the learner; knowing is a being-at-work that is always going on in us, unnoticed until we settle into it out of distraction and disorder (247b 17–18).

ambiguity (**ὁμωνυμία** *homōnymia*) The presence of more than one meaning in a word, sometimes by chance (as in "bark"), but more often by analogy or by derivation from one primary meaning. (See especially *Metaphysics* Book IV, Chapter 2.) A city or society is called healthy by analogy to an animal, a diet by derivation. Derived meanings may have many kinds of relation to the primary meaning, but all point to one thing (*pros hen*). Arrays of this truthful kind of ambiguity reflect causal structures in the world. Book V of the *Metaphysics,* mistakenly called a dictionary, is called by Aristotle the book about things meant in more than one way. Thomas Aquinas uses the word analogy to cover all non-chance ambiguity, but it makes a great difference to Aristotle that the meanings of *good* are unified only by analogy, while those of *being* point to one primary instance.

art (**τέχνη** *technē*) The know-how that permits any kind of skilled making, as by a carpenter or sculptor, or producing, as by a doctor or legislator. The artisan is not "creative"; in nature the form of the thing that comes into being is at work upon it directly, while in art the form is at work upon the soul of the artisan (1032b 14–15). Aristotle agrees with sculptors that Hermes is in the marble, and let out by taking away what obscures his image. Aristotle concludes that the origin of motion that produces statues is the art of sculpture, and incidentally the particular sculptor (195a 3–8). The artwork or artifact has no material cause proper to itself (192b 18–19 —though a saw needs to be of a certain kind of material to hold an edge); in general the artisan uses the potencies of natural materials to counteract one another. The surface of a table strains to fall to earth, but the legs prevent it, while the legs strain to fall over and the tabletop prevents it, and similarly with the roof and walls of a house.

articulation (**λόγος** *logos*) The gathering in speech of the intelligible structure of anything, a combination of analysis and synthesis. A definition is one kind of articulation, but there are many others, including a ratio, a pattern, or reason itself. It can refer to anything that can be put into words—an argument, an account, a discourse, a story—or to the words into which anything is put—a word, a sentence, a chapter, a book. When statements describing the soul are said to be *logoi enuloi* (403a 25), this does not mean that they are "in matter" but that an articulation of some material basis is included in them, the adjective being analogous not to "embodied" but to "ensouled." Translating *logos* as formula is misleading, since it has no implication of being the briefest, or any rigid, formulation of anything. In some translations, the word formula becomes a formula for a rich and varied idea; the word articulation is a slight improvement, used here wherever nothing better was appropriate.

being-at-work (**ἐνέργεια** *energeia*) An ultimate idea, not definable by anything deeper or clearer, but grasped directly from examples, at a glance or by analogy (1048 a 38–39). Activity comes to sight first as motion, but Aristotle's central thought is that all being is being-at-work, and that anything inert would cease to be. The primary sense of the word belongs to activities that are not motions; examples of these are seeing, knowing, and happiness, each understood as an ongoing state that is complete at every instant, but the human being that can experience them is similarly a being-at-work, constituted by metabolism. Since the end and completion of any genuine being is its being-at-work, the meaning of the word converges (1047a 30–31, 1050a 22–24) with that of the following:

being-at-work-staying-itself (**ἐντελέχεια** *entelecheia*) A fusion of the idea of completeness with that of continuity or persistence. Aristotle invents the word by combining **ἐντελές** *enteles* (complete, full-grown) with **ἔχειν** *echein* (= **ἕξις** *hexis,* to be a certain way by the continuing effort of holding on in that condition), while at the same time punning on **ἐνδελέχεια** *endelecheia* (persistence) by inserting **τέλος** *telos* (completion). This is a three-ring circus of a word, at the heart of everything in Aristotle's thinking, including the definition of motion. Its power to carry meaning depends on the working together of all the things Aristotle has packed into it. Some commentators explain it as meaning being-at-an-end, which misses the point entirely, and it is usually translated as "actuality," a word that refers to anything, however trivial, incidental, transient, or static, that happens to be the case, so that everything is lost in translation, just at the spot where understanding could begin. A number of Aristotle's most definitive explanations stand or fall with the meaning conveyed by *entelecheia.*

cause (**αἰτία** *aitia*) The source of responsibility for anything. It thus differs in two ways from its prevalent current sense: in always being a source (1013a 17), rather the nearest agent or instrument that leads to a result, and in referring more to responsibility for a thing's being as it is than for its doing what it does. To understand anything is to know its cause, and such an understanding is always incomplete without an account of all four kinds of responsibility: as material, as form, as origin of motion, and as end or completion (*Physics* Book II, Chapter 3).

chance (**αὐτόματον** *automaton* or **τύχη** *tuchē*) Any incidental cause. At 197b 29–30, Aristotle invents the etymology **τὸ αὐτό μάτην** *to auto matēn,* that which is itself in vain (but produces some other result). Chance events or products always come from the interference of two or more lines of causes; those prior causes always tend toward natural ends or human purposes. Chance is thus derivative from the "teleological" structure of the world, and is the reason nature acts for the most part, rather than always, in the same way. In the *Physics,* chance that is peculiarly relevant to a human being is distinguished as fortune or luck under the name *tuchē,* but in the *Metaphysics* the two senses are merged.

contradictory (**ἀντίφασις** *antiphasis*) One of a pair of opposites which can have nothing between them, such as white and not-white.

contrary (**ἐναντίον** *enantion*) One of a pair of opposites which can have something between them, such as white and black; but the opposition need not be extreme, and could be between two shades of gray.

contemplation (**θεωρία** *theōria*) The being-at-work of the intellect (**νοῦς** *nous*), a thinking that is like seeing, complete at every instant. Our ordinary step-by-step thinking (**διάνοια** *dianoia*) aims at a completion in contemplation, but it also presupposes an implicit contemplative activity that is always present in us unnoticed. To know is not to achieve something new, but to calm down out of the distractions of our native disorder, and settle into the contemplative relation to things that is already ours (247b 17–18). An analogy to the relation between step-by-step reasoning and contemplation may be found in two ways of looking at a painting or a natural scene; one's eyes may first roam from part to part, making connections, but one may also take in the sight whole, drinking it in with the eyes. The intellect similarly becomes most active when it comes to rest.

deprivation (**στέρησις** *sterēsis*) The absence in something of anything it might naturally have. Aristotle regards the distinction between the deprivation, which is opposite to form, and the material, which underlies and tends toward form, as a clarification of and advance over the opinions that came out of Plato's Academy (*Physics* Book I, Chapter 9).

elements (**στοιχεῖα** *stoicheia*) Originally any things in a row, then the letters of written speech, the word finally came to refer to the ultimate kinds of material constituents of things, often considered to be earth, water, air, and fire, with perhaps a fifth celestial element, the ether, added. Aristotle uses the word in this way in discussions of earlier thinkers, but he denies that the simplest kinds of bodies can be elementary, since they can change into one another

(988b 29–34), and argues that in any case no account of elementary constituents can ever explain anything that is genuinely whole (1041b 11–33).

end (τέλος *telos*) The completion toward which anything tends, and for the sake of which it acts. In deliberate action it has the character of purpose, but in natural activity it refers to wholeness. Aristotle does not say that animals, plants, and the cosmos *have* purposes but that they *are* purposes, ends-in-themselves. Whether any of them is in another sense for the sake of anything outside itself is always treated as problematic in the theoretical works (194a 34–36, 415a 26–b3, 415b 15–21, 1072b 1–3), though *Politics* 1257a 15–22 treats all other species as being for the sake of humans. As a settled opinion found throughout his writings, Aristotle's "teleology" is nothing but his claim that all natural beings are self-maintaining wholes.

example (ἐπαγωγή *epagōgē*) The perceptible particular, in which the intelligible universal is always evident. The word induction, which refers to a generalization from many examples, does not catch Aristotle's meaning, which is a "being brought face-to-face with" the universal present in each single example. A famous simile in the last chapter of the *Posterior Analytics* (100a 12–13) is often taken to mean that the universal must be built up out of particulars, just as a new position of a routed army is built up when many men have taken stands, but it means just the opposite: it only takes one man to take a stand, after which every other soldier, down to the original coward, will be identical to him. The rout corresponds to the condition of someone who has not yet experienced some universal in any of its instances. Evidence for this interpretation is found in many places, such as *Posterior Analytics* 71a 7–9 and *Physics* 247b 5–7, in which Aristotle unmistakably says that one particular is sufficient to make the universal known. That in turn is because the same form that is at work holding together the perceived thing is also at work on the soul of the perceiver (*On the Soul* 424a 18–19).

first philosophy (**πρῶτη φιλοσοφία** *prōtē philosophia*) The study of immovable being, or of the sources and causes of all being. Aristotle's organized collection of writings on this topic was called by librarians **τὰ μετὰ τὰ φυσικά** *ta meta ta phusika,* "what comes after the study of natural things," but neither this phrase nor any like it is ever used by Aristotle. He names the topic in the order of the things themselves, rather than from the way we approach and think about them, and calls physics second philosophy. What we call metaphysics, the post-natural, is for Aristotle the pre-natural, the source and foundation of motion and change. That form is present in all things is a starting-point for physics; what form is must be clarified by first philosophy (192a 34–36).

form (**μορφή** *morphē* or **εἶδος** *eidos* or **ἰδέα** *idea*) Being-at-work (1050b 1–2). The failure to understand form as being-at-work has led to hopeless confusion about many of Aristotle's claims, for example that sense-perception is a reception of form without material (424a 17–19). It is often said that Aristotle imports the form/material distinction from the realm of art and imposes it upon nature. In fact it is deduced in *Physics* Book I, Chapter 7 as the necessary condition of any change or becoming. In a compressed way in *Physics* Book II, Chapter 1, and more fully in *Metaphysics* Book VIII, Chapter 2, it is argued that arrangement is insufficient to account for form, which is evident only in the being-at-work of a thing. *Morphē* never means mere shape, but shapeliness, which implies the act of shaping, and *eidos,* after Plato has molded its use, is never the mere look of a thing, but its invisible look, seen only in speech (193a 31). *Idea,* from the same root as *eidos,* is used primarily when technical discussions within Plato's Academy are referred to, but the English words "idea" and "ideal" are distortions of it, suggesting something that can only be present in thought, which no one who used the Greek word intended.

genus (**γένος** *genos*) A divisible kind or class. It might arise from arbitrary acts of classification, in contrast to the *eidos* or species, the kind that exactly corresponds to the form that makes a thing just what it is. The highest general classes are the so-called "categories," the irreducibly many ways of attributing being. *Metaphysics* Book V, Chapter 7 lists eight of these: what something is, of what sort it is, how much it is, to what it is related, what it does, what is done to it, where it is, and when it is. *Categories* Book IV adds two more: in what position it is, and in what condition.

harmony (**ἁρμονία** *harmonia*) Originally a fitting or joining of parts in carpentry, by extension a tuning or attunement of high and low sounds to form a musical scale or a melody, and finally the ratio of any things mixed together (408a 5–9). It was a popular ancient idea that the soul is a harmony of the body, and many commentators in our time have said that this is very close to Aristotle's claim that the soul is a being-at-work of the body, but Aristotle dismisses the comparison of the soul to a harmony as absurd (408a 13). The proportions of body parts or ratios of body fluids in living things are not responsible for life but are themselves produced by the formative activity of soul; the metaphor of fitting does not fit the work of the soul (408a 5).

imagination (**φαντασία** *phantasia*) A power of the soul that perceives appearances when perceptible things are absent and thinks without distinguishing universals (429a 4–8, 434a 5–11). The imagination is identified in *On Memory and Recollection* as the primary perceptive power of the soul (449b 31–450a 15). Thus, many activities discovered in *On the Soul* may be collected and attributed to the imagination, such as perceiving common and incidental objects of the senses, being aware that we are perceiving, discriminating among the objects of the different senses (425a 14–b 25), distinguishing flesh or water (429b 10–18), and perceiving time (433b 7). Also, implicit within the power of imagination to behold images (*phantasmata*), there must be imagination in a second sense, *eikasia*, by which we can see an image as an image (*eikōn*) or likeness (*On Memory and Recollection* 450b 12–27).

impasse (**ἀπορία** *aporia*) A logical stalemate that seems to make a question unanswerable. In fact, it is the impasses that reveal what the genuine questions are. Zeno's paradoxes are spectacular examples, resolved by Aristotle's definition of motion. In *Metaphysics* Book III, a collection of impasses in first philosophy, Aristotle writes "those who inquire without first being at an impasse are like people who do not know which way they need to walk" (995a 36–38). The word is often translated as "difficulty" or "perplexity," which are much too weak; it is only the inability to get past an impasse with one's initial presuppositions that forces the revision of a whole way of looking at things.

incidental (**κατὰ συμβεβηκός** *kata sumbebēkos*) Belonging to or happening to a thing not as a consequence of what it is. The word "accidental" is appropriate to some, but not all, incidental things; it is not accidental that the housebuilder is a flute player, but it is incidental. To any thing, an infinity of incidental attributes belongs, and this opens the door to chance (196b 23–29).

intellect (**νοῦς** *nous*) That in the soul, as well as separate from the soul, which thinks. It has both a broad and a narrow sense; see thinking.

lead back (**ἀνάγειν** *anagein*) To produce an explanation while leaving the thing explained intact. Aristotle leads back all motion to change of place without reducing all motion to change of place.

material (**ὕλη** *hulē*) That which underlies the form of any particular thing. Unlike what we mean by "matter," material has no properties of its own, but is only a potency straining toward some form (192a 18–19). Bricks and lumber are material for a house, but have identities only because they are also forms for earth and water. The simplest bodies must have an underlying material that is not bodily (214a 13–16).

motion (**κίνησις** *kinēsis*) The being-at-work-staying-itself of a potency *as* a potency (*Physics* Book III, Chapters 1–3). Any thing is the being-at-work-staying-itself of a potency as material for that thing, but so long as that potency is at-work-staying-itself as a *potency,* there is motion (1048b 8–9). Motion is coextensive with, but not synonymous with change (**μεταβολή** *metabolē*). It has four irreducible kinds, with respect to thinghood, quality, quantity, and place. The last named is the primary kind of motion but involves the least change, so that the list is in ascending order of motions but descending order of changes.

mover (**κινοῦν** *kinoun*) Whatever causes motion in something else. The phrase "efficient cause" is nowhere in Aristotle's writings, and is highly misleading; it implies that the cause of every motion is a push or a pull. In *Physics* Book

VII, Chapter 2, it is argued that in one way all motions lead back to pushes or pulls, but this is only a step in a long argument that concludes that every motion depends on a first mover that is motionless (258b 4–5), and the only kind of external mover that is included among the four kinds of cause in *Physics* Book II, Chapter 3 is the first origin of motion (194b 29–30). That there should be incidental, intermediate links by which motions are passed along when things bump explains nothing. That motion should originate in something motionless is only puzzling if one assumes that what is motionless must be inert; the motionless sources of motion to which Aristotle refers are fully at-work, and in their activity there is no motion because their being-at-work is complete at every instant (257b 9).

nature (**φύσις** *phusis*) The internal activity that makes anything what it is. The ideas of birth and growth, buried in the Latin origins of our word, are close to the surface of the Greek word, sprouting into all its uses. Nature is evident primarily in living things, but is present in everything nonliving as well, since it all participates in the single organized whole of the cosmos (1040b 5–10). Everything there is comes from nature, since all chance events and products result from the incidental interaction of two or more prior lines of causes, stemming from the goal-seeking activities of natural beings, and all artful making by human beings must borrow its material from natural things.

now (**νῦν** *nun*) The indivisible limit of a time. The word "moment" is not a suitable translation, because it refers to a *stretch* of a continuous process, nor is the word "instant" appropriate, since the now is relative to a soul that can recognize it. Time arises from the measurement of motion, which can only take place in a soul that can relate two motions by linking them to a now (223a 21–26).

number (**ἀριθμός** *arithmos*) Any multitude, whether of perceptible things, definite intelligible things, or empty units. The last named, the pure numbers of mathematics, Aristotle calls the numbers *by* which we count (219b 8), but the word normally refers to the first kind, the numbers which we count, such as the dozen eggs in a carton, a multitude *of* something. The remaining kind of number is alluded to at 206b 30–33; Plato seems to have taught that higher and lower forms are not related as genus and species but in the same way as a number and its units. That is, the unity of wisdom, courage, temperance, and justice, for example, would not be a common element contained in them all, but the sum of them all as virtue, the eidetic number four. A number in any of its senses is something discrete and countable, and never includes continuous magnitude; it therefore excludes fractions, irrationals, negatives, and all the other things brought under the idea of number when Descartes fused the ideas of multitude and magnitude, or of how-many and how-much, into one. It is thus a paradox, lost on us, when Aristotle says that time is both continuous and a number, but only in resolving that paradox is it possible to see how he understands time.

of-this-kind (**ποιόν** *poion*) Being of one or another sort is a more direct and immediate feature of things than having a quality (**ποιότης** *poiotēs*), a word that Aristotle rarely uses.

place (**τόπος** *topos*) The stable surroundings in which certain kinds of beings can sustain themselves, in which alone they can be at rest and fully active. When *dis*placed, anything strives to regain its appropriate place. This idea of place depends on the prior idea of the cosmos as an organized whole, in which there is no void. The contrary idea of space, as empty, homogeneous, and infinite, Aristotle regards as an abuse of mathematical abstraction: the positing of an extension of body without body.

potency (**δύναμις** *dunamis*) The innate tendency of anything to be at work in ways characteristic of the kind of thing it is; the way of being that belongs to material (1050a 18). The word has a secondary sense of mere logical possibility, applying to whatever admits of being true (1019b 31–33), but this is never the way Aristotle uses it. A potency in its proper sense will always emerge into activity, when the proper conditions are present and nothing prevents it (1047b 35–1048a 21).

primary (**πρώτη, πρῶτον** *prōtē, prōton*) First in responsibility. It is translated as first when it means first in time.

rest (**ἠρεμία** *ēremia* or **στάσις** *stasis*) Motionlessness in whatever is naturally capable of motion (202a 4–5). A natural being at rest is still active. Nothing is inert.

sense perception (**αἴσθησις** *aisthēsis*) Always the reception of organized wholes. Never sensation as meant by Hume or Kant, as the reception of isolated sense data. Though the proper objects of sense perception are the colors, sounds, smells, tastes, and feels about which the distinct senses can never be in error (418a 7–15), perceiving is always the reception of forms (424a 17–19), each of which is the being-at-work of some independent thing as a whole (1050b 1–2). Hence, while one can be mistaken about what the perceived thing is incidentally, for instance that it is the son of Diares (418a 16, 20–21), that perceived thing is taken in with an intelligible structure, what it is for that thing to be, which is not a construct of ours but, as a form without material, is present to the contemplative intellect and no more subject to error than are the senses themselves (430b 28–30).

separate (**χωριστόν** *chōriston*) Able to hang together as a whole, intact, on its own. Aristotle never uses the word to mean "separable." Mathematical things are not separate, not because they happen never to be found in isolation, but because they do not compose anything that could be at work. By the same token, the form *is* separate (1017b 27–28 and 1029a 27–30). When the form is remembered or reconstructed in thought as a universal, it is separate only in speech or articulation, but the form as it is in itself, as a being-at-work and a cause of being, is separate simply (1042a 32–33). In a number of places, such as 193b 4–5, Aristotle says that form is not separate except in speech, but this is always a first dialectical step, articulating the way form first comes to sight; at 194b 9–15 he already balances it with the opposite opinion, and points to the inquiry in which the question is resolved.

so much (**ποσόν** *poson*) Not isolated quantity but the muchness or manyness that belongs to something. The former is studied by the mathematician; the latter is present in nature.

source (**ἀρχή** *archē*) A ruling beginning. It can refer to the starting point of reasoning, but the usual translations "principle" or "first principle" are rarely adequate, since the word most often refers to a being rather than to a proposition or rule. The divine intellect deduced in Book XII of the *Metaphysics* is not an explanatory principle but a being on which all other beings depend.

thinghood (**οὐσία** *ousia*) The way of being that belongs to anything which has attributes but is not an attribute of anything, which is also separate and a *this* (1028b 38–39, 1029a 27–28). Whatever has being in this way is an independent thing. In ordinary speech the word means wealth or inalienable property, the inherited estate that cannot be taken away from one who is born with it. Punning on its connection with the participle of the verb "to be," Plato appropriates the word (as at *Meno* 72B) to mean the very being of something, in respect to which all instances of it are exactly alike. Aristotle elaborates this meaning into a distinction between the thinghood of a thing and the array of attributes—qualities, quantities, relations, places, times, actions, and ways of being acted upon—that can belong to it fleetingly, incidentally, derivatively, and in common with things of other kinds. He concludes that thinghood is not reducible to any sum of attributes (1038b 25–28, 1039a 1–3). It thus denotes a fullness of being and self-sufficiency which the Christian thinker Augustine did not believe could be present in a created thing (*City of God* Book XII, Chapter 2); he concluded that, while *ousia* meant *essentia,* things in the world possess only deficient kinds of being. *Substantia,* the capacity to have predicates, became the standard word in the subsequent Latin tradition for the being of things. A blind persistence in this tradition gave us "substance" as the translation of a word that it was conceived as negating.

thinking (**νοεῖν, νόησις** *noein, noēsis*) This is Aristotle's broadest word for thinking of any kind, from the contemplative act that merges with the things it thinks (429b 3–7, 430a 19–20, 431b 17), through all the ways of dividing up and putting back together those intelligible wholes (430b 1–4), to mere imagining (427a 27–28); but it is also used in its most governing sense for the primary kind of thinking that underlies them all (430a 25), as a synonym for contemplation (*theōria*). Throughout *On the Soul,* Aristotle uses at least two dozen different words for the various kinds of thinking. (See the note preceding Book III.) Modern philosophers such as Descartes and Locke homogenize the objects of all these activities into contents of consciousness or "ideas in the mind." Since Aristotle does not do this, but keeps the kinds of thinking distinct as one primary and many derivative capacities for being-at-work, the word "mind" does not correspond to anything in his vocabulary. It is used here once (427a 26) in a quotation from Homer for a crude notion of thinking that Aristotle criticizes.

thinking things through (**διάνοια** *dianoia*) The step-by-step thinking involved in all reasoning (*logismos*) as opposed to contemplation, and in all conceiving (*hupolēpsis*) as opposed to imagining.

this (**τόδε τι** *tode ti*) That which comes forth to meet perception as a ready-made, independent whole. A *this* is something that can be pointed at, because it holds together as separate from its surroundings, and need not be constructed or construed out of constituent data, but stands out from a background. The mistranslation "this somewhat" reads the phrase backward, and is flatly ruled out by many passages, such as 1038b 25–28.

underlying thing (**ὑποκείμενον** *hupokeimenon*) That in which anything inheres. It can be of various kinds. Change presupposes something that persists. Attributes belong to some whole that it not just their sum. Form works on some material. An independent thing is an underlying thing in the first two ways, but not in the third (1029a 20–28).

universal (**καθόλου** *katholou*) Any general idea, common property, or one-applied-to-many. It is never separate and can have no causal responsibility, unlike the form, which is a being-at-work present in things, making them what they are (1040b 28–30, 1041a 4–5).

virtue (**ἀρετή** *aretē*) Any of the excellences of the human soul, primarily wisdom, courage, moderation, and justice. Though they depend on learning or habituation, Aristotle regards them as belonging to our nature. Without them we are like houses without roofs, not fully what we are (246a 17–246 b3).

what it is for something to be (**τί ἦν εἶναι** *ti ēn einai*) What anything keeps on being, in order to be at all. The phrase expands **τί ἐστι** *ti esti,* what something is, the generalized answer to the question Socrates asks about anything important: "What is it?" Aristotle replaces the bare "is" with a progressive form (in the past, but with no temporal sense, since only in the past tense can the progressive aspect be made unambiguous) plus an infinitive of purpose. The progressive signifies the continuity of being-at-work, while the infinitive signifies the being-something or independence that is thereby achieved. The progressive rules out what is transitory in a thing, and therefore not necessary to it; the infinitive rules out what is partial or universal in a thing, and therefore not sufficient to make it be. The learnèd word "essence" contains nothing of Aristotle's simplicity or power.

Bibliographic Note

Brief, excellent introductions to Aristotle's thinking as a whole, to his work as a biologist, and to his account of the contemplative intellect may be found, respectively, in lectures by Jacob Klein, "Aristotle: an Introduction," Linda Wiener, "Of Lice and Men: Aristotle's Biological Treatises," and William O'Grady, "About Human Knowing." Klein's lecture is in his collected *Lectures and Essays* (St. John's College Press, 1985), Wiener's lecture appeared in *The St. John's Review*, Vol. XL, No. 1 (1990-91), and O'Grady's is in his collected writings in *The St. John's Review*, Winter, 1986, all of which are available from the St. John's College Bookstore, P. O. Box 2800, Annapolis, Maryland 21404. Klein's lecture was also anthologized in the book *Ancients and Moderns* (Basic Books, 1964), which might be found in some libraries. All three write from foundations of solid learning, but address all interested and thoughtful readers, rather than some group of professional scholars.

Martin Heidegger's interpretations of crucial passages in Aristotle's *Metaphysics*, *Nicomachean Ethics*, and *Physics* may be found, respectively, in his recently compiled books *Aristotle's Metaphysics Θ 1-3* (Indiana University Press, 1995), *Plato's Sophist* (Indiana University Press, 1997), and *Pathmarks* (Cambridge University Press, 1998). Sympathetic commentary on Aristotle often takes its bearings from within a medieval Latin scholarly tradition, while antagonistic commentary is often rooted in beliefs that come out of a scientific tradition that began in the seventeenth century. Heidegger's work bypassed both these traditions, in an effort that was at once poetic and deeply philosophic, to read Aristotle directly. The loss of the comfortable correctness one may achieve within a tradition of interpretation is the price one pays for a chance at genuine insight. Even if one cannot follow Heidegger's lead in all respects, he brings Aristotle's thinking powerfully to life. And his renderings of Greek words, which often appear far-fetched on first reading, always turn out on deeper study to be well founded and worth thinking about. His assertions make an excellent beginning for dialectic.

Outstanding examples of products of the Latin tradition at its best may be found in Joseph Owens's book *The Doctrine of Being in the Aristotelian Metaphysics* (Pontifical Institute of Mediæval Studies, Toronto, 1951), and Yves Simon's *The Great Dialogue of Nature and Space* (Magi Books, 1970). Owens's book is the outstanding scholarly work on Aristotle produced in the twentieth century, and bears directly in many ways on the reading of *On the Soul*. Simon's book is a transcribed course of informal lectures that sets out in a clear and elementary way the opposed views of the world in Aristotle's *Physics* and Descartes's writings; it too prepares the way for reading *On the Soul*.

The current standard secondary literature about *On the Soul* is of higher quality than that which deals with most of Aristotle's other writings. This is true also for the *Nicomachean Ethics*, because many of those who write about these two books believe that they might learn something from Aristotle, rather than simply put him in his place. The recent anthology *Essays on Aristotle's De Anima*, edited by Martha Nussbaum and Amélie Rorty (Oxford, Clarendon Press, 1992), is one of a number of similar volumes that deal with various books and topics in Aristotelian scholarship, and it is distinctly superior to the others. Following an introductory section, the volume leads off with an essay the author of which confidently proclaims that no one nowadays could possibly agree with anything Aristotle says about the body or soul. Many of the subsequent essays take up the challenge and prove him wrong. Various interpretations are represented, and anyone whose interest is aroused may follow them up in many directions through the other works that are cited.

There is another sort of literature, though, that seems to me to be more truly descended from *On the Soul.* The phrase philosophical biology has been applied to the writings of a number of authors in this century. Erwin Straus was a psychiatrist whose splendid books, *The Primary World of Senses* (The Free Press of Glencoe, 1963) and *Phenomenological Psychology* (Basic Books, 1966), attempt the Aristotelian task of articulating the relation between the sensing, moving, and thinking being and the true world of experience, which is by no means the world constructed by mathematical physics. Adolf Portmann was a zoologist who did not reduce animal lives to the Darwinian struggle for reproductive success, but saw that the preservation of any species must be for the sake of some kind of self-expression in which the animal finds completion. His books include *Animals as Social Beings* (Viking, 1961) and *Animal Forms and Patterns* (Schocken, 1967). Marjorie Grene, who has written a good introductory book about Aristotle, emphasizing biological themes, *A Portrait of Aristotle* (University of Chicago Press, 1963), has also written essays on these and other philosophical biologists, which are included in her book *The Understanding of Nature* (Reidel, 1974). There are many other writers whose work falls in this broad area that begins with the study of living beings and continues into philosophic reflection. Two extraordinary neurologists are among them: Kurt Goldstein, whose 1939 book *The Organism* has been recently republished (Zone Books, 1995), and Oliver Sacks, whose philosophic talent is evident in *Awakenings* (Dutton Obelisk Books, 1983) and several other books.

Index

About Joe Sachs

photo by John Bildahl

Joe Sachs has taught for twenty five years at St. John's College, Annapolis, Maryland, where from 1990 to 1992 he held the N.E.H. Chair in Ancient Thought. His other translations of Aristotle include *Physics* (Rutgers University Press), *Metaphysics* (Green Lion Press), and *Nicomachean Ethics* (R. Pullins Company).